"IT DEPENDS"

Why learning is *Not* guaranteed

"IT DEPENDS"

"Coach, will I score better after instruction?" **"It Depends!"**

"Professor, can I get an 'A' in your course?" **"It Depends!"**

"Business training, will it help?" **"It Depends!"**

By Michael Hebron[1]
with Stephen Yazulla, Ph.D.[2]

[1]PGA of America Hall of Fame
[2]Professor Emeritus, Neurobiology and Behavior

IT DEPENDS!

For information about this title or to order other books and/or electronic media, contact the publisher:

Learning Golf, Inc.
495 Landing Avenue, Smithtown, Long Island NY 11787
www.michaelhebron.com

ISBNs:
978-1-937069-11-7 (hardcover)
978-1-937069-12-4 (softcover)

First Edition Published 2023

Printed in the United States of America by LightningSource, Inc.
This book is Sustainable Forestry Initiative ® SFI® certified. www.LightningSource.com

Editorial Assistance: Nannette Poillon McCoy
Cover Design by Kirsten Hebron
Interior design: 1106 Design

Library of Congress Cataloging-In-Publishing Data
Library of Congress Control Number

Hebron, Michael, PGA Hall of Fame with Yazulla, Stephen, Ph.D.
It Depends! 1st ed.
Includes bibliography references

With the help of award-winning educators and scientists, Michael has been researching the brain's connection to learning for over two decades. He attended Harvard University's Connecting Mind-Brain to Education Institute, Dr. Eric Jensen's Teaching with the Brain in Mind workshops and had the honor of spending time with renowned Dr. Robert Bjork at the UCLA Learning Lab.

Michael has been invited to talk about the topic of the brain's connection to learning and instruction at seminars around the world, and to professional organizations, businesses, Universities and Colleges including Yale and MIT, authored six books and hundreds of articles and been a guest on TV shows including the Golf Channel, Charlie Rose Show, and NBC's Today Show.

Michael organized the first National PGA of America Teaching and Coaching Summit, the first PGA of Europe and first Canadian PGA Teaching and Coaching Summits. A member of World of Golf Teachers Hall of Fame, and PGA of America Hall of Fame, and a recipient of PGA National Teacher of the Year and National PGA Horton Smith Awards.

Organization Memberships have included Learning and the Brain Society (LBS), American Society of Training and Development (ASTD), American College of Sports Medicine (ACSM)

National Center for Science Education (NCSE), Association for Scholastic Curriculum Development (ASCD), Phi Delta Kappa International Education Association (PDKIEA), PGA, USGA.

Stephen Yazulla, Ph.D was on the faculty of Stony Brook University, Stony Brook, NY from 1974–2013. He is Professor Emeritus in the Department of Neurobiology and Behavior, and Department of Ophthalmology. The Retinal Research Laboratory of Dr. Yazulla was devoted to the neurophysiology, neuroanatomy, and neurochemistry of the retina for 35 years, while funded by the National Eye Institute.

In excess of 100 publications document his research findings. He has numerous invited presentations at national and international universities and presentations at Scientific Conferences. Some of his Societal Memberships include: Association for Research in Vision and Ophthalmology, International Brain Research Organization, International Cannabinoid Research Society, International Society for Eye Research, and Society for Neuroscience.

Dr. Yazulla's teaching experience includes Undergraduate and Graduate courses in Biology, Physiology, and Neuroscience; Pre-Nursing Laboratory courses; and lectures in the Ophthalmology Residency Program.

Table of Contents

Preface

"I cannot teach anybody anything;
I can only make them think"
—Socrates

> "What are the most effective ways to approach learning and performing any topic"? PGA Hall of Fame instructor, Michael Hebron and university Professor Dr. Stephen Yazulla, Ph.D. discuss this question here. In their view, **IT DEPENDS** on severable variables that go beyond just providing information and delves into the internal factors that influence students. It is also the only answer to questions like: How much? When? Where? What direction? In life and learning, often the answer is **IT DEPENDS.**

Hebron

I was a young teacher with little experience when I began coaching but had a strong desire to grow and was determined to learn all I could about the golf swing. Over the next 20 years or so I invested resources in books, videos, cameras, seminars, traveling to observe other teachers, and attending many educational programs that were about the golf *swing*. Then years later, I added research about the brain's connection to the nature of learning and that process of change to my instructional-tool box. I moved on from just providing golf-swing information to becoming more informed about what did and did not support student learning.

My coaching experiences have included hands-on individual golf instruction and classes, presentations and workshops at Universities and professional conferences in over twenty countries and the lesson is not what I say, but what the students take away.

I see coaching sessions not as a time for students to get something right, but as an opportunity for them to experiment mentally and physically. I see my time with students as providing information for them to think about and then act on. This approach is similar to a lab investigation that causes learning to come alive and connect with real world experience and options.

When it comes to coaching, in my opinion, there are a lot of effective men and women helping and guiding their students. Each may have their way of supporting growth and development of skills, which can be a different way than other coaches. Every student is different, so is every coach. And perhaps flexible thinking is the most effective coaching tool. Keep in mind what the great John Jacobs said, "*there is some good in every golf book; it just depends on who's reading it.*"

Anything I, or any coach, may share with students is most likely available at a click of a button on the internet. However, when it comes to learning, it is not just the information that makes the main difference, but how it is shared with students. I ask all students if they want to experience instruction that will try to teach them, or instruction that will help them learn. There is a difference. I have never had a student say they just wanted me to teach them. **Keep in mind that when you teach, you can win or lose, but when you help someone to learn, you always win.**

Yazulla

My educational background is in Psychology, Cellular and Systems Neuroscience. My professional academic/research experience started at Stony Brook University in 1974. My teaching over the next 40 years ranged from lectures to classes from five students to a thousand students. Subject matter included most aspects of systems physiology and neuroscience to undergraduate and graduate students.

Lecturing to the students was only part of it. Office hours to review material on upcoming exams, grading exams and final grades, with career aspirations in the balance, course, and major advising, writing letters of recommendation for advanced study were all part of being an academic faculty. This varied educational experience exposed me to interacting with students with wide abilities and motivations on individual bases as well as in much larger groups. There also were invited lectures and presentations at universities and numerous international conferences that expanded opportunities for me to convey information effectively to peers as well as to students.

Hebron

When I brought up the topic of **IT DEPENDS!** for a book to Dr. Yazulla, a Neurobiology and Behavior professor, he agreed. When it comes to experiencing learning, it depends on conditions that go beyond the information provided. Oddly, neither of us could recall a book on this topic. After his guidance with our book "*Learning with the Brain in Mind*" I looked forward to teaming up with Dr. Yazulla again. So off we went with the aim of writing one that would give readers the opportunity to recalibrate the value of an **IT-DEPENDS** view during learning opportunities. In a world filled with so many "right or that is wrong" opinions about instruction, the **IT DEPENDS** view often gets overlooked. The obvious contingencies include motivation, effort, past experience, expectations, manner of instruction, among others. More recently, the effect of stress and emotions on learning and performance is receiving increased attention from several fields of science and from me.

Yazulla

Our collaboration started several years ago as informal conversations. I had played golf at the Smithtown Landing Golf Course for many years. I had a copy of the book "*Golf Swings, Secrets and Lies*" by Michael Hebron, who happened to be the Master PGA professional at Smithtown Landing Golf Course. I learned from "*Golf Digest*" that Michael Hebron was routinely listed in the *Top 50 Golf Teachers* in the country, and more recently, had been inducted into the PGA Hall of Fame. Although I played the course many times, I had never met him. I thought to go up to the course one day and ask Michael to autograph the book for me. I introduced myself and asked for his autograph. He graciously obliged and asked me what I did. "I am a Neurobiology Professor at Stony Brook University." This answer was the start of many conversations about the brain's role in learning. Michael had been working on a topic about Brain-Compatible Learning in terms of an emotionally safe teaching-learning environment and how it can help instructors to be aware of how the brain affected learning and performance.

We had many conversations on what constituted "Brain-Compatible Learning," eventually publishing a book entitled "*Learning with the Brain in Mind*." Our experiences have much in common, but with enough differences to result in very interesting give-and-take. My role was to provide the brain-science context on learning environments that Michael had been researching for two decades. Then Michael floated the idea of a concept wrapped around "*It depends*." The germ of this idea was our consistent response to a question we often received from students, in the line of "Will I succeed?" The answer of course is "*It depends*." The more we talked, it became clear that an It-Depends topic provided us an opportunity and framework to integrate our very different teaching experiences that combined, exceeds 100 years.

Rather than focus on the instructors and "external environment" as it affects student learning and performance, attention here in **IT DEPENDS!** is given to the internal features of motivation, strategies, experience, and particularly stress and emotion as they affect learning and performance. We have had many opportunities to observe how these factors affect students. Much of students' success depends on what they "bring to the table," regardless of the instructor or environment involved. The answer then to the question of "will I succeed" many times falls on the student. *It depends!*

> *"Whatever is received is received*
> *according to the manner of the receiver."*
> —St. Thomas Aquinas

Hebron

Dr. Yazulla and I have the same aim as other teachers and coaches. We want the information we share to be learned, remembered, and put to use effectively by our students. We propose that any instructor becomes more efficient when they become informed about what can support or suppress acts of deep learning. There are numerous environments in which learning is the aim, but is learning taking place? *It depends!*

Unlike Dr. Yazulla, I am not a neuroscientist. The thoughts I share here are based on the evolution of coaching methods from many years as a professional golf instructor. Also, I have been directly and indirectly influenced by interactions with educators at the Center for Education Research and Innovation, Queensland Brain Institute, UCLA Learning Lab, Harvard University Mind-Brain Education Institute, Johns Hopkins University Neurology and Cognitive Science Depts. Some of what I have included here comes directly from their research, which I share because others may not be aware of this information that caused me to rethink how I now spend time with students.

Yazulla

Our ideas and suggestions often use the game of golf as an example. Any instructional activity, sports, or academia could be substituted for "Golf." Our goal is to improve learning and performance for students by reducing stress in coaching, teaching, and performing environments. My experience in academia certainly differs from a coaching environment. However, I have been on the other side of teaching and coaching, as a student from grammar school through graduate school and as a player in organized baseball, bowling, golf, rifle team and karate. This it-depends perspective makes me appreciate the approach to instruction that Michael has been promoting for so many years. I am happy to be part of this effort.

Hebron

This book cannot possibly replace personal interactions between providers and receivers of information. What each individual believes about teaching and learning is based on what they already know and fortunately, that which is known can be increased. With that in mind, let's move along with the aim that readers may find some different and useful ways to look at instruction, learning and performing.

There are numerous publications from Cellular, Systems and Behavioral Neuroscience devoted to these issues as they relate to learning and performance. Some of these will be alluded to and cited appropriately. A general overview to the **Brain and Nervous System** that will present the more technical vocabulary used throughout can be found in the **Appendix**.

Introduction

"How we learn is different than how we think we learn."
—Dr. Robert Bjork, UCLA Learning and Forgetting Lab

In the previous book, *Learning with the Brain in Mind*, we presented the case that the methods and approaches to learning come before learning and will affect how well students learn and perform. The attention here is more on how students and instructors best incorporate information and translate that information into acts of memory and performance. Such a supportive environment is "Brain Compatible" with how students learn.

Hebron

A major point here is the more that providers of information are informed about the nature of learning, the more prepared they are for employing methods that support positive growth and development. When approaches to learning enhance learning potential, performance potential also improves. A master of anything was first a master at learning. Some approaches to instruction enhance learning potential; some do not. *It depends*!

Studies from Harvard University and other research centers have documented that the design and structure of "the approach" to learning influence a student's pace of progress in sports and school, their future employment opportunities, living conditions, physical and mental well-being, and progress with other endeavors. Some approaches to learning are more compatible with the brain's connection to the nature of learning than others.

Why mention the brain—because there is a critical LINK between our brain, human development, and deep learning. All of our conscious and subconscious behavior is either controlled by or modulated by the brain. By harnessing this LINK, we open possibilities for improving methods of instruction and training instead of teaching without learning. Ideally, during learning opportunities, instructors and students are drawn together as a team with overlapping interests and aims.

While the information and personal beliefs of teachers and golf coaches can vary on the same topic, all want their students to experience learning that leads to successful performance. The student's aim is to learn; the educator's aim also should be to use methods for sharing information that support this aim of students. A brain-compatible approach to instruction guides the development of a student's ability to think critically, learn personal skills and apply them in situations beyond in which they were first learned. We think it is important for instructors to appreciate the internal factors affecting the students' motivation, stress, etc. in order to successfully engage the students in any lesson.

Yazulla

Rather than discuss learning theories that highlight the instructors and learning environment, *It Depends!* discusses conditions affecting the mental and emotional states of students as they relate to learning and performance. Our brain should not be seen as just a "Black Box" to be filled at will without having to consider any negative effects that teaching methods and learning environments have on our brain. However, the ideal is not reality. Stress, fatigue, past experience, expectations are among the numerous factors that can affect the brain during learning. Understanding the source of these factors can mitigate negative effects on learning. Performance, related to but different than learning, is similarly affected (see Chapter Three).

Instruction in academia is quite different from sports or athletic instruction, as with the environment in which learning and performance occur. Perhaps a major difference is the relative amount of mental and physical skills that are involved in any activity. Mental skills can require minimal physical activity whereas all sports activity requires mental skills. This topic is treated in Chapter Four.

> *"The secret in education lies in respecting the student"*
> —Ralph Waldo Emerson

Hebron

Providing information during instruction is similar to planting seeds. After planting, will healthy plants be the outcome? It depends not only on conditions of preparation and care that go into just planting seeds. Success also depends on the planter's knowledge of what is required to care for the seeds you are planting, i.e., sunlight, water, pH, fertilizer, etc. Likewise in learning environments, providing information is just one side of the coin. Influences on learning go beyond just providing information, and like the seeds, depend on the care and preparation devoted to acts of learning as well as knowledge of factors affecting the students.

Approaches to deep learning create meaningful connections including student to topic, provider to receiver of information, emotions to environment, past experience to new experience, poor outcomes to workable outcomes, translate information to meaning; all are based on efficient communication among brain neurons throughout the brain. What approaches support learning? *It depends!* Strategies of drill, play, focus and diversify are treated in Chapters Seven, Eight, and Nine.

During the late 20th and early 21st century the topic of learning underwent several important developments. The aim was to create adaptive approaches to learning to be used creatively by students and instructors in ever-changing and different situations. These studies suggested:

1. a learner-centered approach to growth and development

2. founded on the social nature of learning

3. influenced by emotions and past experience, and

4. individual student differences

This approach is highlighted by constructive, self-regulated learning that is collaborative and sensitive to the context.

Keep in mind that each student is different: different backgrounds, different physically, different strengths, weaknesses, different information base, different emotions, etc. We could go on with differences but there is a common denominator, "the student's capacity to achieve." Yes, students are different but their "capacity to achieve" will be supported or suppressed by where and how new information is provided. It depends largely on the motivation that is driving a student to learn and perform (Chapter Five).

Teaching and learning are not synonymous. Some individuals learn at an acceptable pace while others struggle and do not. How does instruction provided become information and skills learned? How do we go from not knowing to knowing? It depends on pedagogy, the methods and approaches used during learning.

Yazulla

"Pedagogy" is derived from two Greek words meaning "a leader of children." Its original use referred to slaves in ancient Greece who took boys to school and waited for them. Pedagogy, since the 16th century, refers to the art, science or profession of teaching that takes into consideration the interactions between instructors, students, and information during learning.

For centuries, pedagogy consisted of a top-down approach with a teacher determining the content and manner by which information was conveyed to a largely passive student group. This style is referred to as teacher-centered. However, the current pedagogical approach is student-centered with students taking an active role in the learning process.

Hebron

Some methods of teaching react to where students are, and not to where they are going. On the other hand, learn, develop and growth approaches to learning are less reactive and concerned with poor outcomes. What are referred to as "grade" or "outcome" environments can have a limited view of the nature of learning. **They look for what's right or wrong and not for what is being learned.** They are teaching to get something right and overlook the educational value of struggling with unintended outcomes. Confusion and struggle are welcome additions to the nature of learning in emotionally-safe learning environments. This topic is treated in Chapter Six.

The design of the approach to learning has a direct influence on a student's emotions, memory, self-image, self-evaluation, among other considerations. Emotion is an element of everything humans do. Students often travel between positive or negative emotions depending on their experience in a learning environment.

Yazulla

Our goal is to improve learning and performance for students by reducing stress in the coaching, teaching, and performing environments (Chapter Two). Generally speaking, effective learning starts with information, moves onto memory, then to the ability to recall and apply. What follows is a series of chapters that relate primarily to major conditions, internal and external to the student and instructor that affect student learning and performance. There is no one size to fit all students, instructors, or context; there are nuances in each of the conditions that are discussed. We treat:

- ○ Opening Insights by Michael Hebron

- ○ the nature of learning and our teaching experiences (Chapter One)

- ○ the positive and negative effects of stress and emotion (Chapter Two)

- ○ how is learning determined? Inferred from performance (Chapter Three)

- ○ motor learning differs from conceptual learning (Chapter Four)

○ types of motivation and effects on learning and performance (Chapter Five)

○ the value loaded word "Failure" and the role of "Errors" (Chapter Six)

○ the relative values of "drilling" versus "deliberate play" (Chapter Seven)

○ whether or not "expert models" are useful (Chapter Eight)

○ whether it is better to "Focus" or "Diversify" as learning strategies (Chapter Nine)

○ practical suggestions to implement Brain-Compatible Learning (Chapter Ten)

"The foundation of every state is the education of its youth"
—Diogenes

Opening Insights

Hebron

The aim of this book is not to fix what is not working, but rather to offer insights into the nature of learning. We suggest that these insights support deep learning, effective memory, and performance. To like, to want, and to need are often seen as different emotions, but when it comes to the nature of learning, these terms apply. We like to learn; we want to learn, and we need to learn. This book connects these emotions with the aim that students will profit from what instructors are sharing.

Neuroscience has succeeded in unraveling activity within critical chemical and electrical pathways used for forming memory and creating movement. These processes are influenced by our past experience, prior knowledge, and emotions. Subject content is merely the skeleton of learning, not its flesh. Here we suggest ways to positively leverage the brain's information processing skills. How do we do this? *It depends*! At times, instruction can be intellectually interesting, but educationally vacant. *It depends*!

In the fast paced, result-oriented culture that we live in today, some approaches to learning have become more a product of just providing content, rather than placed in context for a more critical understanding. This occurs not only in schools, but at times in the sports instruction industry.

PLAYFUL = **P**owerful **L**earning **A**bout **Y**ourself **F**inds **U**seful **L**earning

Our brain acts as an information observatory: comparing, combining, and clarifying while creating changes that lead to learning. Make the methods of instruction brain-compatible first; then focus on subject content. A playful instructional process, one free of negative judgments, is the aim here. Note, that the root of the term "SCHOOL" is from the Greek word *Skholé*, meaning "LEISURE."

"We don't stop playing because we grow old; we grow old because we stop playing."
—George Bernard Shaw

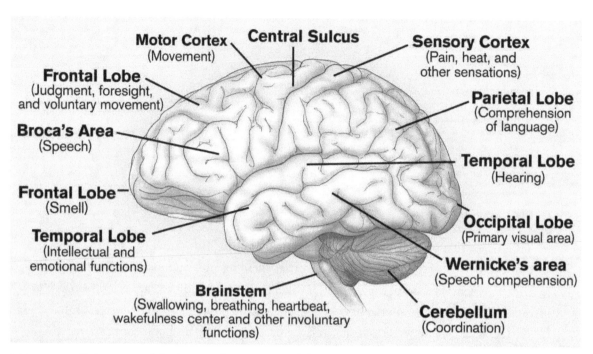

FIGURE 1: *Schematic of the human brain, viewed from the left side, illustrating generalized regions of function.*

Experiencing deep learning is not about the information we learn and can repeat; it is more about how we think and what we can do with that information. Cognitive growth is more likely to be supported when students are required to explain, elaborate, or define information. Striving to produce an explanation often forces the learner to integrate and elaborate knowledge in new ways.

At times when students are not learning it may not be the students who need more attention, rather the method of instruction may need to be reevaluated to be consistent with the nature of learning and memory. Teachers and coaches, who understand the content of the course as well as the nature of learning, overcome some of the factors that negatively influence each student's learning progress.

The same situation exists in every school from grammar school through college. At the end of the semester, there are A, B, C, D and F students, the same teacher and subject, but different results. Why? *It depends*!

Brain-based, student-centered approaches to learning simultaneously interact with a student's past and present thoughts, with an eye to the future. It is a process that understands that the methods used to provide and gather information during learning opportunities are important.

Student-centered approaches are supported by a flexible and portable process that does not have a precomposed order. This process is not filled with many details taking up space in our psyches. It provides the opportunity for personal insights that are supported with general non-specific concepts that support neurodevelopment. Neurodevelopment systems include attention control, memory, motor, sequential, spatial, higher thinking, social thinking, and language. Factors that influence neurodevelopment system include education, health, peers, values, family, environment, genetics, and emotions (Levine, 2002).

Deep Learning Is Active Learning

During deep learning, changes take place within the millions of cell connections within the brain's information highway. One aim of student-centered learning is to keep students interested, which leads to enjoyment and fun during trial-and-error activities of a self-discovery process. This is why there is no formula for how we do things.

An it-depends view uses approaches that constantly respond to the value found in both workable and unintended outcomes. Students learn that answers to "What," "Why," and "When" questions are influenced within the context of their experience. The quality of learning depends, for better or worse, by internal and external factors regardless of subject topic and content.

If one existed, a "Dictionary of Learning" would not include expanded explanations of most topics. Incomplete information could evoke curiosity and ingenuity in students as they tried alternative ways to solve a problem. It-depends views and answers are filled with imagination and internal self-communication during a personal journey of change.

A brain-based view recognizes that unintended outcomes are just moments of transition to more workable results. From the UCLA LEARNING LAB, "Struggle is one of the most useful learning tools." Some approaches to learning treat students as if they are playing some version of dodge ball, trying to avoid anything that may be unintended, instead of a mindset of learning what to do differently to change that outcome.

Are teaching and learning synonymous? No. How do we go from not-knowing to knowing? It depends on the methods and approaches used during learning. A term, familiar to many is pedagogy.

When students learn and develop their capacities, an effective pedagogical approach not only offers information, but it also recognizes the various needs of students while adjusting to the surrounding conditions. Effective pedagogy supports effective teaching and learning.

I have never seen myself as a perfect teacher, but I have learned that there are perfect learners—they are called human beings. Learning is a natural skill we all enter the world with, actually learning while in our mother's womb. Unfortunately, this skill can be unintentionally suppressed by some approaches to the instruction of any topic.

A student's aim is to learn; an educator's aim is to use methods for sharing information that support the aim of students. A brain-compatible approach to instruction guides the development of a student's ability to think critically, learn personal skills and apply them beyond situations in which they were first learned.

The Human brain is often compared to a computer; but they have very different operating systems.

- One has been programmed, the other self-gathers information.

- One has to be perfect to work at all; the other is naturally flexible and tolerant of unwanted outcomes.

- One has a central processor; the other has no one central control, it is free to respond.

- One stores information exactly; the other does not remember exactly the information it is presented. It only recalls important relationships, no details.

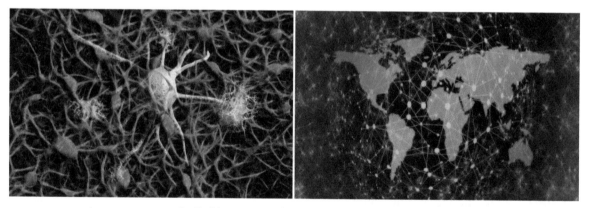

FIGURE 2: *A nerve network (left) compared to a world-wide web network (right). Both are extremely complicated, but the nerve network occupies the relatively small space in the brain. With its many billions of interconnections, the nerve networks are continuously being remodeled based on experience.*

My aim is to have approaches to instruction that are:

○ **Smart S**tudent's **M**inds **A**re **R**eally **T**alented

○ **Safe S**tudents **A**lways **F**irst **E**nvironment

○ **Playful P**owerful **L**earning **A**bout **Y**ourself **F**inds **U**seful **L**earning

Brook Adams of the Boston School Board stated back in 1879 that knowing that students cannot be taught everything, it is best to teach them how to learn.

Today, by combining golf-swing information with instructional methods that are compatible with the nature of learning, I found that students learn faster and retain skills longer. Students are less frustrated in these learn-and-develop environments, than in a teach-and-fix to-get-it right environment that I had been using in the past.

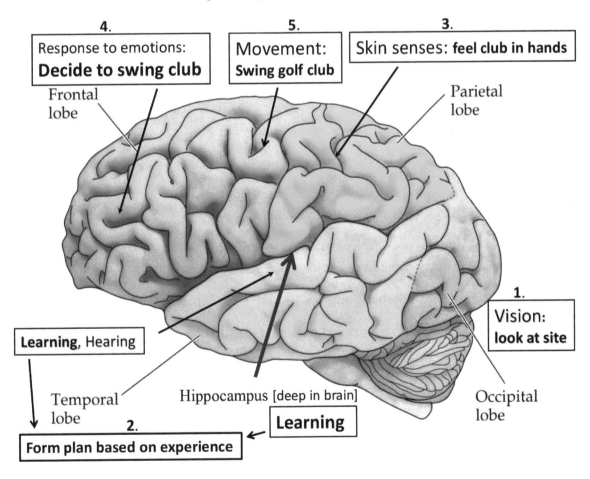

FIGURE 3: *The brain on golf, illustrating a sequence of events leading to a swing.*

In the past, like today there was no shortage of information and opinions about the mechanics of the golf swing, but little if any conversation about:

○ the nature of learning

○ the brain's connection to learning

○ how emotions and mindset influence learning, the nature of memory, and recalling information

Slow to Learn

I had some blind spots about teaching and learning during the first half of my career. But as Will Durant, a famous historian suggested, "My education is the progressive realization of my own ignorance." And I was ignorant. During my first 20 years of coaching, I was slow to learn that the student can be a real educator, providing information for both themselves and their instructor. Back then I was being influenced by traditional customs and was just providing golf-swing information during instruction. Perhaps some readers who are teachers of any topic will identify with the following:

○ I was—Slow to learn that having preconceived ideas about students and having the answer for them can be damaging.

○ I was—In charge—not a collaborator.

○ I was—Giving answers—overlooked the value of 1) self-discovery, b) poor unintended outcomes, and c) what students already know.

○ I was—Pointing out poor habits and failures—and did not recognize that what some call failure, actually is valuable feedback for future reference points.

○ I was—Trying to fix unwanted outcomes—and not helping to change a student's insights.

○ I was—Just reacting to unintended outcomes—and not providing a proactive student-centered learning experience.

○ I was—Overlooking that the unwanted outcome is over, there is nothing to fix. Pay attention to what to do next, not what to fix.

○ I was—Trying to teach a subject—not supporting a journey of playful self-learning.

○ I was—Trying to teach details using expert models—and not using a learning model consisting of general, non-specific just in-the-ballpark concepts.

- I was—Giving commands—without providing choices. Do you like A, B, or C. I did not recognize a choice as one of the most powerful tools we can give students.

- I was—Slow to realize that a lesson is an opportunity to experiment, not just the time to try to get-it-right.

- I was—Slow to understand what I was saying is not the lesson, but what students take away from instruction is their lesson.

- I was—Trying to improve mistakes—not helping to enhance the students' approach to learning. By improving methods of learning you can improve performance.

Brain-compatible approaches to deep learning create meaningful connections including:

- student to topic

- provider to receiver of information

- emotions to environment

- past experience to new experience

- poor outcomes to workable outcomes

- translates information to meaning.

I have been asked why I say progress during instruction is more about the student than the coach or content of the lesson. For me, it is very straightforward. While coaches are important, every student is different. Some have more time to train, some have more experience, some have more talent, some have played other sports, some are in different physical condition than others, some have more emotional control, some have more motivation, some handle the ups and downs of the game and training results differently than others. The list goes on, and we also coach many individuals, but only a few make lasting progress, and very few excel. Why, because students are all different and it depends on how teachers in schools and coaches in sports adjust to these differences.

Your Notes

CHAPTER ONE

The Nature of Learning

Hebron

Over time I would come to the conclusion that learning environments are different from teaching environments.

In teaching environments, the instructor provides answers.

In learning environments, the student is guided in the direction of developing the tools to self-discover their answers.

Modern science has researched the nature of learning with the combined efforts of Social Sciences, Behavioral Research, Developmental Psychology, Neuroscience, and other fields of research. They have uncovered insights about experiencing deep learning that are often overlooked. Experiencing deep learning is not just supported by a straight-line journey of providing information, it is also a process filled with a variety of twist-and-turn adjustments. Knowing is a combination of understanding and applying. The nature of learning would be a terrible gift to misuse.

Hebron—My Coaching Experience with The Nature of Learning

When I began coaching golf, I traveled to observe other teachers and attended many educational programs that were filled with opinions about the golf swing. I eventually became a member of the PGA of America and certified Master Professional. Later in my career when some of my students were not reaching what I thought was their potential, I asked myself what was I overlooking? What was missing?

Those questions caused a personal and career changing "Ah Ha Moment." I would come to realize the golf-swing information I was providing was only one side of the instructional story. The other and just as important side is how does the instructional method support student learning, and then recalling that information after instruction? Back then, I was overlooking and was uninformed about the topics of learning and memory, which was a large oversight that I would soon change.

This reality created a conflict for me. About the same time that I was becoming aware of what studies were uncovering about the nature of learning, I started to receive some recognition from the golf industry for my work as a teacher.

The conflict? I was being recognized as an award-winning teacher but knew little about the brain, the gateway to learning, and how students best learn the information I was providing. For many years I may have been subject-matter rich, but I was also nature-of-learning poor; hence—The conflict.

With help from leading researchers from Harvard, UCLA, MIT, Stanford, Vanderbilt, Columbia, and many other research centers, I began to recognize that some methods of instruction are more capable of providing support for learning, while others unintentionally suppress the quality of learning students can experience. I started to understand that deep learning is an active process, not a passive one by which students are just given information. That reality motivated me to become more informed about brain-based, student-centered approaches to teaching and learning.

Over the past twenty-five plus years, I became more informed about how teaching and learning can be affected in negative and positive ways by influences including: emotion, motivation, past experience, stress, memory, expectation, environment, the student's ability, the approach in use, focus, failure, success, self-image, self-control, training, practice, self-assessment, judgment and criticism from others and ourselves. I learned that all these variables influence learning and performing, and emotions have the strongest impact.

The huge and important influence that "emotions" have on learning and developing is an insight I was made aware of as a student at Harvard University's Connecting Mind-Brain to Education Institute. At that time, this reality was often side-stepped. Today this topic is taken more seriously into consideration and is addressed here further in Chapter Two.

Yazulla—My Teaching Experience with The Nature of Learning

Looking back at my professional teaching career starting in 1974, I realized it was better to cover less material well, than race over a lot of information superficially. The major constraints are time and what students think they need to advance to the next stage. Passive learning situations ordinarily involve a Lecturer and Listeners. The Listeners are not totally passive. During any lecture or conference presentation, a variety of sensory and motor activities do take place among individuals that can determine how much information is retained. The following is a brief survey of my academic teaching environment in a classroom.

Note: Laboratory courses, Performing Arts and the Trades are different in that active student participation is essential.

Classroom settings are quite varied: ranging from all grades of elementary, high school, college, graduate, and medical schools as well as technical, adult education, self-help symposia, and so on. In all these cases there is usually one person who directs discussion or provides the information to the group. The students in these settings are largely passive participants until such time to allow for questions and answers. The presentations are either oral or a combined audio/visual. The participants can look, listen, additionally take notes, or record the session. This is referred to as Teacher-Centered-Instruction.

Taking notes is problematic because no one can focus on two things at once. When you are writing, there is an editing process by which you filter what is relevant while writing it down. Remember, what you are writing down or typing on your laptop has already been spoken, and you are missing most of what is being said as you write. Given my frustration trying to take notes as a student, as an instructor then, I tried to be sensitive to this disconnect between lecture speed and the ability to listen while taking notes.

The manner of presenting class material changed drastically over my academic lifetime (1974–2015). Class attendance ranged from small seminars of 3–6 people to large groups of 1000 or more. In a large lecture hall, I was on a stage with a throat microphone, two projection screens behind me, and a sea of students in front and above, in a balcony. In the early years, my classes were smaller (25–40 students) and presentations involved writing on a blackboard with chalk; sometimes supplemented with handouts.

The beauty of chalk and blackboard is that the lecturer cannot write faster on a blackboard than they can talk, which means that the student-note taker has a chance to keep up with the material being presented. Later on, overhead projectors and slides were used so that larger groups could see material from the back of the room. The downside of projectors allowed for more information to be presented because the lecturer was freed from writing as they spoke spoke.

The result was that many students used tape recorders to record the lecture for later playback. These students did not take notes during the lecture; rather they listened to the lecture and then to the tape later on. Of course, during the lecture, some students would ask for clarification on a topic. These other students then had an instant-replay not only of the lecture, but also of the questions and answers. Note-takers were at a disadvantage and would need to collaborate with other students to fill in gaps. Collaboration among students was excellent but did not compensate for incomplete notes.

Eventually, as technology improved, I used PowerPoint™ presentations—computer-generated slides with pictures and text projected onto a large screen. The disadvantage of Power Point™

was that even more information could be put onto a single slide, faster than the students could write or even read. The adjustment was to simplify the slides and slow down the lecture. As technology improved, the lecture slides were uploaded on the University website from which students could log in and print the slides prior to the lecture. The next step was to video-record the lecture on the University website that then could be viewed on-demand on a computer or listened to on an audio-podcast. As predicted, with the availability of the lecture material before the lecture and a telecast of the lecture afterward, attendance in class dropped by over 50%.

Over a course of 40 years there was a progression from talking while writing on a blackboard, with the students following as best as they could, to watching reruns of the lectures on university podcasts or broadcast TV. The result of all this progress was that students seemed to depend only on the lecture handouts of the PowerPoint™ presentation, rather than using the textbook to flesh out the lecture notes. It seemed that the more effort that was made to make the material accessible to the students, the less effort the students made to delve into the details of the material. Overall, student/faculty interactions were reduced: during class, after class, even at office hours; except for the day before an exam—sad.

Finally, with the advent of off-site class locations and the difficulty of staffing summer classes, lectures were simulcast to remote locations, and then rebroadcast as online courses–without faculty involvement. This displacement in space and time puts the burden of learning squarely on the student. There is no more interaction of the give-and-take of ideas as occurs in small classes or seminars. The internet has put endless information at our fingertips, but without critical presentation, analyses, or discussion. So, in a university setting, in which the dispensing of course material became more automated and less interactive, the opportunities for discussion and growth were more limited than the old-fashioned 'chalk and a blackboard.' The irony is that with the Covid-19 restrictions, such online-learning in 2020–2021 became the normal condition.

My academic experience as a University Professor has given me quite an overview of the educational system as it changed over the years. Overall, students are the same. However, the culture they grew up in, technology, career aspirations and social norms have changed drastically from when I grew up and throughout my teaching career. Since I retired, students are more computer savvy and proficient at typing now than writing. In fact, cursive writing is becoming a lost art, giving way to printing and texting. As I look back on my academic career, what worked, what did not work, what would I change in retrospect? This topic of *It-depends!* has allowed me to look back with a new perspective that differs from "This is what you need to do—Listen and Learn."

Hebron

Knowledge Of

There is knowledge of THAT, knowledge of HOW and knowledge of WHY. Some approaches to learning are designed to move students beyond gaining just the knowledge of THAT and onto to gaining knowledge of HOW and WHY that is learned, remembered and used.

When organizing approaches to instruction that support effective learning, there are several conditions to take into consideration including: Prepare us for complete living. WHERE information is shared, WHO is sharing it, WHEN it is shared and, perhaps most importantly, the emotional STATE of the students and individuals sharing the information.

The design of the approach to learning has a direct influence on a student's self-image, self-evaluation, among other considerations. For better or worse, during learning opportunities, people often bounce between positive and negative thinking that the learning environments create. Will deep learning take place? *It depends.*

GAPS

There are misconceptions about approaches to learning that can cause gaps between a student's skills and the level of achievement that students experience. Insights offered here suggest these **GAPS** (**G**one **A**re **P**ossible **S**olutions) can be addressed by rethinking instructional methods that are used during instruction.

Education cannot be given to students, but rather is gained by students who experience learning environments that develop their conceptual understanding to solve problems creatively. Information that tries to help someone is not as useful as information that helps someone help themselves.

More Insights

Painter's History of Education, written 136 years ago in 1886, is filled with some valuable insights that are paraphrased below.

> The science of education is incomplete. The principles which should influence educational methods are to be found in human nature. This truth, which long remained unnoticed or inoperative, has been emphasized by the educational reformers of *modern* times. (ital. mine; modern times in 1886!)

The great law underlying physical and mental development of man is a SELF-activity. Every truly educated man is self-made. Learning by the consequences of trial-and-error is the greatest thing preparing us for complete living.

Sound education is symbolized by a tree. A little seed, which contains the design of the tree, is placed in the soil. It now germinates and expands into trunk, branches, leaves, flowers, or fruit. The whole tree is an uninterrupted chain of organic parts, a plant which existed in its seed. Man is similar to the tree. Within the newborn are 'faculties' which unfold during life and form themselves gradually into a whole.

The process of physical and mental growth is assisted and guided by the function of education, a realization of possibilities.

Develop the anchors of the mind with the natural laws of learning.

Learning and teaching should be the introduction into active life.

NOTE: All these insights were suggested over 130 years ago.

Deep Learning

It's fair to say that approaches to learning should support a student's ability to experience deep learning. This is what Dr. Dee Fink Ph.D., Director Instructional Developmental Program, University of Oklahoma calls, "significant learning experiences" and others, "life-long learning" or "Deep Learning."

Today the interest in the topic of education has expanded to the point in which insights about the ways we humans actually learn are now available. This information is moving students beyond just surface learning into deeper learning.

Making mistakes can be very valuable. I call them "Desirable Developmental Difficulties." Cognitive science confirms that unworkable outcomes are a biochemical necessity of Deep Learning. Trial-and-error struggling is the main tool of Deeper Learning.

Questions to students that support Deep Learning include:

    ○ Does what you just learned remind you of anything you already know?

    ○ How does this new knowledge differ/conflict with your prior beliefs about this topic?

○ What was the most difficult thing to grasp during the lesson?

○ What was the easiest thing to grasp?

○ Have you noticed any pattern for what is easy or what is hard to learn for you?

Students should be encouraged to keep handwritten notes; they support deeper learning more effectively than typed notes. The sensory-motor involvement in the brain for writing is greater than for typing.

Students' attention span may only be about 10 minutes or so, after which they start to drift. To adjust, provide information in a variety of ways, and in brief segments.

New knowledge is not created; rather new learning is encoded by integrating new information with prior knowledge.

View Deep Learning as information that has been developed to the point where it can be used in several different environments, beyond the one it was learned in.

Using massed practice of a skill is a popular stratagem for learning, but it is also among the least productive. Variety in training is the most productive way to experience Deep Learning.

Also, mixing different but similar kinds of skills in the same learning session beats the practice of just one topic or skill. For example, baseball players learned to hit curveballs better when the practice sessions were filled with fastballs, change ups, and curve balls, than when they faced only curve balls.

From *The Cambridge Handbook of Thinking and Reasoning* (Holyoak & Morrison, 2013)**,** "Knowledge does not have to grow as a set of detailed events, just in some organized general fashion" (p. 373). The science of learning suggests using general concepts and avoiding details. Learning happens in a network of general concepts, not specific details. One example, the elements of balance, timing, rhythm, alignment, size, shape, etc. develop into one concept, or one act of motion, or one feel in our brain.

One of America's first great thinkers about learning, and founder of progressive education in the United States, John Dewey, built on Francis Parker's insights from 1875 about learning. Dewey saw traditional education as being isolated from reality and passive in its methods. The approaches to learning that Dewey created at the University of Chicago had three principles.

1. Instruction must focus on the development of the student's mind, not on blocks of subject matter.

2. Instruction must be integrated into project-oriented approaches.

3. The progression through years of school education must go from practical experiences to formal information, to integrated studies.

John Dewey's principles required students to make their own observations and predictions, thus developing the student's own thinking skills. In terms of designing efficient learning environments Dewey left us some key ideas:

- The student is the center of learning.

- Learning is an active engagement, with an environment structured for education (not test scores).

- Integration of mind and action, head and hand, academic and vocation promotes learning.

- Learning takes place in context (i.e., real world situations).

- Learning is guided, providing structure for connections between every experience.

- Learning in context gives students insights about the role of information in problem solving.

How?

According to the National Assessment of Education Progress, the manner in which many students are being taught information and skills does not line up with what research has uncovered about

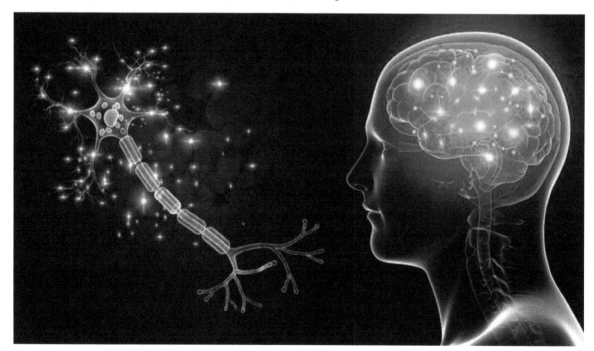

FIGURE 4: *A neuron is the signaling cell in our central and peripheral nervous systems. A typical motor neuron in the spinal cord is illustrated on the left. This type of neuron contacts skeletal muscle.*

how people actually learn. A study into teacher preparation found that, after interviewing more than 100 deans and faculty members as part of a study into teacher-preparation programs, most of them could not explain basic principles about how people learn.

Unfortunately, according to the National Council on Teacher Quality, research on components of learning often is not presented in teacher-preparation programs. Students then suffer if they are not instructed with methods and approaches known to support the brain's connection to learning.

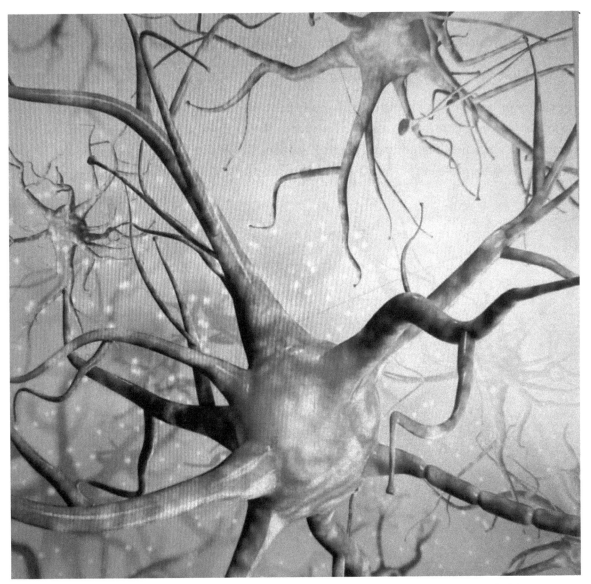

FIGURE 5: *Neurons in the brain have a vast variety of shapes with numerous branch-like processes, extending from a central cell body.*

Neurons

Scientists estimate that there are about 100 billion neurons in our brain. Biologically, a modern human does not differ that much from man of 40,000 years ago. Over time, each generation did not have to start from scratch; they learned from their ancestors over the generations in ever-changing environments.

In general, there are three functional classes of neurons:

1. Ones that detect environmental changes (sensory systems)

2. Ones that store and process incoming information (central nervous system)

3. Ones that activate muscles (motor systems)

All classes of neuron (sample pictured) participate in our survival. Neurons, grouped into their various structures, control the way we interact with our environment. All input to the brain comes from our senses (sight, touch, hearing, pain, etc.). Final coding and integration of all this information takes place in the brain, which then directs the motor system how to respond to all

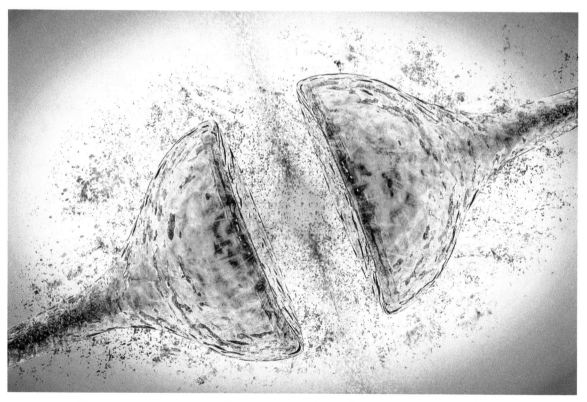

FIGURE 6: *From neuron to neuron, information is exchanged at synaptic endings of neurons by chemical messengers, as illustrated here.*

the incoming sensory information. Learning, perhaps the most important function of the nervous system, is subject to many external and internal influences. Taking these factors into account during instruction will result in learning environments that are brain-compatible. More detail about the structure and function of neurons can be found in the Appendix.

From Vanderbilt University's John Brandsford, *Understanding the Brain: The Birth of a Learning Science*, (2007, p. 93), "Neurobiological research is now informing the design of effective instruction, thereby increasing the probability of interventions being effective, helping educators and coaches design instruction with powerful consequences.

Train Your Brain to Learn

Thanks to advances in neuroscience, talent development professionals, for example, The Association for Talent Development (ATD), have an opportunity to improve approaches to teaching and coaching when current research is put to use. A trainer, coach, teacher, student, consultant can be more effective than in the past with their target audience.

While humans have always looked for ways to improve performance, only recently have we been able to see how the brain operates, receives, learns, and encodes information for future retrieval. An overview of practical suggestions about talent-development professionals includes:

- o Lead by the natural laws of human development, individuals can gradually become their own self-reliant educator. (I am guiding students in the direction of being their own best coach.)

- o Deep learning arrives mostly not by what students are told, but by what they find out for themselves and through practical trial-and-error use, and not merely from books and lectures.

- o Learning strategies are tools that promote gradual development, not perfection, which does not exist. Consistency does not exist, (we cannot even write our own name the same way each time.)

- o There is a valuable culture of development in which inconsistency is an opportunity for growth. (The unintended outcome instructs us for what to do differently.)

- o Have flexible knowledge and portable skills; there is no one way. (There are many ways to play.)

- o No one is broken in need of fixing—the unwanted outcomes are over and can't be fixed. Think what to do next, not what to fix on a journey of development. (Positive learning)

- o The level of a performance is never at one skill level—it is always in a range. (We never move exactly the same way every time.)

○ Deep learning involves revisiting memories, reconstructing them, and then joining them with current experiences. New learning requires reference points based on past experience and prior knowledge, (if this, then that.)

Myths

Many long-held myths about how we learn can be more attractive than the informed realities of 21st century research. Some myths about learning discussed at Harvard's Connecting the Mind-Brain to Education Institute and The UCLA Learning Lab when I was a student there include:

Myth: Basics must be learned so well that they become second nature.
Fact: Over-learning basics at the start can stifle creativity and individual expression.

Myth: Delaying gratification is important.
Fact: Keeping a growing interest and joy in learning leads to more learning.

Myth: There are always right and wrong answers.
Fact: Correctness depends on context.

Myth: Good learners know what is out there.
Fact: Life-long learners are not know-it-alls.

Myth: Forgetting is a problem.
Fact: Memory often prevents new or novel learning and use beyond personal biases.

Myth: Memorization is necessary.
Fact: When possible, relating new information to personal experience is better than memorization.

Myth: There is a limit to what can be learned.
Fact: It appears there is no limit to storing information. Unfortunately, our ability to recall is limited. Therefore, approaches to learning must take care they develop and support the skill of recalling information. Use metaphors and stories.

Myth: Orderly learning is the aim.
Fact: Students need unpredictable environments to gain understanding. Struggling creates access to new learning for the long term. "Order does not establish new learning that lasts" Dr. Robert Bjork.

Myth: We use only 10% of our brain.
Fact: Most, if not all of the brain is engaged in learning and everything else, we do.

Myth: We are aware of what we are learning.

Fact: We often learn without actually meaning to do so. A large percentage of learning is below our conscious awareness.

Myth: Learning is putting new information into the brain.
Fact: Deep Learning actually requires and depends on prior knowledge and assumptions that connect with new information.

Myth: Learning is based on getting it right.
Fact: Uncertainty, confusion and struggles support learning.

Myth: Long sessions on one thing are more useful than several short sessions.
Fact: Several short sessions are more useful than one long session.

Myth: Study or train in one location during one session.
Fact: Vary and change locations when training or studying at home.

Yazulla—Myth of Learning Styles

Myth. Learning Styles—It is common practice today that people consider themselves to be primarily visual or auditory learners. That is, they think they learn more quickly when information is presented to them by sight or hearing. This idea was formalized into a hypothesis and is referred to as *Preferred Learning Styles* based education. The simplest categorization of learners is VAK (Visual, Auditory, and Kinesthetic Learners). There is no doubt that people have such a preference, but do these groups actually learn and perform better when information is presented in their preferred manner? Do visual learners perform better when presented with visual information, and do auditory learners perform better when presented the same information by sound? Although *Learning Styles* is accepted by the vast majority of educators in Europe and the United States, is this a valid way to group and teach students? Is *Preferred Learning Styles* based education effective and does it result in better learning?

Fact. Despite the persistence and acceptance of *Learning Styles*, there is **NO** empirical evidence that teaching to preferred-learning styles is effective. For a reviews of this controversy, (see for example, Pashler et al., 2009; Newton, 2015; Willingham et al., 2015; Aslaksen & Loras, 2018; Furey, 2020). **Rigorously controlled** studies in adults and fifth-graders showed no benefit when teaching students to their preferred-learning style (Rogowsky et al., 2015; 2020). To be clear, there was no difference in learning (i.e., performance) in auditory learners regardless of whether the information was presented in a visual or auditory format. The same was true for visual learners; it did not matter whether the information was presented in a visual or auditory format. Having information presented in the preferred-learning style **did not** result in better learning in comprehension and performance.

However, the most provocative finding was that visual learners performed better than auditory learners on measures of listening and reading comprehension.

To repeat, overall, Visual Learners performed better than Auditory Learners on the same task, regardless of how the information was presented.

This was found for adults as well as for fifth-graders (Rogowsky et al., 2015; 2020). The implication for K-12 education is clear regarding auditory learners, to quote:

> "This would suggest that to achieve superior comprehension, which is vital for classroom learning, all students need as much opportunity as possible to build strong reading skills. Thus, *contrary to the learning-style hypothesis* (ital. mine), it may be particularly important to focus on strengthening reading skills in all students, especially for auditory learners" (Rogowsky et al., 2020, p. 6.)

Learning Styles is not an evidence-based approach to education. Though accepted by many, perhaps by some intuition, there are no valid studies to support this approach. Indeed, *Learning Styles* has been called a 'neuromyth' and referred to as the "educational equivalent of homeopathy: a medical concept for which no evidence exists, yet in which belief and use persists" (Newton, 2015.)

Why do people believe in an educational approach that has no valid research to support it? Perhaps, because it fits a general assumption, or others believe it, or it is perceived to be common knowledge; or because of extensive media exposure promoting it. In any event, if you want students to see something, hear something, or feel something, teach for that outcome, not to a preferred learning style.

Hebron—Going Forward

As with the issue of *Learning Styles*, it is important for educators, coaches, etc. to become more familiar with research on learning. It really does not take much effort to stay up-to-date.

Give students choices, demonstrate A, then B, and C. Then have them choose which one they like and based on their choice, move forward. If they make a choice that does not support progress, help them change their current insights and concepts.

Have students make mistakes during instruction. For example, in golf ask them to flip their hands when swinging through impact. Then ask them to do it differently, without showing them how. It helps students learn the feel of what to do when students are aware of what mistakes feel and look like.

During learning opportunities, change the context every 5 to 10 minutes. For example, when coaching the topic of alignment, spend 5 minutes on a green, 5 minutes in a bunker, 5 minutes on a tee, and 5 minutes in the fairway.

When the methods and approaches to learning are compatible with the Nature of Learning some of the qualities include:

- ❍ Are emotionally supportive
- ❍ Are physically safe
- ❍ Are developmentally appropriate
- ❍ Give students choices
- ❍ Avoid judgments
- ❍ Are free of negative criticism
- ❍ Find unworkable outcomes useful for improving
- ❍ Change poor insights
- ❍ Introduce information or skills that are just beyond current skill levels

Gaining an Education

It is often said that every child has the right to receive a good education, but an education is gained and cannot be given. Every child should have the opportunity to develop the tools to gain an education.

When gaining an education, all these activities are organized in the brain.

- ❍ An ongoing process in the development of new views and attitudes
- ❍ A continuing process of rethinking past experience and prior information
- ❍ A mental reorganization that opens possibilities of future activity
- ❍ An internal change that supports external growth
- ❍ A process that precedes constructive activity
- ❍ A joining of content with application in context

○ A transformation of information into flexible knowledge and portable skills

○ A process that supports the manipulation of what is known

"A brain-informed approach to learning addresses key issues for the educational community that can open new pathways and improve research, policies, and practices."
—Bransford, 2007.

There is an inventory of prior workable and unworkable outcomes encoded in our brain that act as the ship's captain, sub-consciously guiding our mind on an active journey through life's ever-changing environments.

The brain is primed for action before the mind decides to move. There is a 'readiness potential' in the brain that precedes a conscious decision to move by about a half a second. Our past experiences and emotions sub-consciously guide this process.

Readiness Potential exists during approaches to learning that recognize how emotions and past experiences influence one's ability to learn, recall and apply. Unfortunately, our conscious mind often will try to justify overlooking the value of readiness potential in favor of consciously focusing on details.

Feedback

Giving feedback is an important element of instruction but is often seen completely backwards. A positive impact on a student is more than just information boiled down to a set of prescribed steps. "For many, feedback is about telling people what we think is wrong with their performance. The research is clear: telling people what we think of their performance doesn't help them thrive and excel. It actually hinders learning" (Buckingham & Goodall, 2016). They went on; "put attention on what is done well and certainly not on what someone other than a student think is done poorly."

Coaches, who structure their conversation with students as a dialogue, in which all ideas count, are more effective than those who do not. Also moving the conversation off the student and on to the topic being explored supports learning. Coaching conversations have been found to be more effective when they are non-judgmental.

Outcomes are what they are. In golf, they are high or low, slow or fast, left, right etc. and should not be seen as failure. It was the only outcome possible for that swing. The outcome is not judged or criticized when coaching supports learning. Admitting what the outcome actually is, for better or worse, supports learning.

Outcomes are either wanted or unintended, and the unintended results point to what could be done differently. Every outcome is perfect. The student did everything correct for that outcome. If you want a different outcome, help students learn to do something differently using what was done in the past as a reference point. Saying that went long or right perpetuates learning, saying that it was a bad shot does not.

Yazulla—Importance of Sleep

Sleep is a daily period in which normal sensory, motor, and mental activity is reduced. Such a period of relative quiescence occurs throughout the animal kingdom. Insects, fish, birds, mammals, all have daily episodes of sleep; many during the night and others during the day. Even plants shut down during regular periods in the day. Most people know that during sleep the body and brain rejuvenate and repair themselves. Indeed, sleep deprivation has severe effects on the mental and physical states of the body, both of which are quite interconnected.

A sleep cycle lasts about 90 minutes and goes through five stages, including light sleep (stages 1, 2), deep sleep (stages 3,4) and dream sleep (stage 5). Dreaming is accompanied by rapid-eye-movements and is commonly referred to as REM sleep. During REM sleep, brain-wave activity resembles one who is awake and alert, and breathing, heart rate and blood pressure rise. During deep sleep, heart rate, body temperature, breathing, and muscle tension decreases. Unlike REM sleep, deep sleep is characterized by slow brain waves, and is also referred to a slow wave sleep (SWS). It is largely during deep SWS that tissue and body repair occur, and previously-learned events are replayed in the brain.

Research over the last 20 years or so has shown that events learned during the day are replayed in the brain during deep SWS sleep. Regardless of whether the learning was by vision, hearing, touch, or movement, they are replayed during SWS sleep in those areas of the brain involved in the learning task (see for example, Huber et al., 2004; Abel et al., 2013; Michon et al., 2019; Rubin et al., 2022).

Although deep SWS occupies only about 15% of a sleep cycle, SWS is very important for the consolidation of memory. Studies show that disruption of SWS impairs recall of novel and rewarded learned events, rather than familiar or neutral events (Giradeau et al., 2009; van de Ven et al., 2016; Michon et al., 2019). These studies suggest the importance of a varied and positive environment during learning. Because of the novelties, these learned events are more rehearsed by the brain during SWS. In the following chapters, as we discuss the numerous factors that contribute to and interfere with learning and performing, the role the sleep will be addressed throughout.

Hebron

Why do I coach with the nature of learning in mind? Here are five reasons: 1) someone asked for help, 2) pass on what I have learned, 3) make students think, 4) I may learn something, and 5) I may rethink something.

For me, coaching is an intervention that can change or add value to what students think about. I recognize that my approach to coaching matters because it happens before and then influences positively or negatively acts of learning in my student.

I see my coaching approach as a path **of** learning, not a path **to** learning. It is coaching shaped by its purpose, which is to help students learn from their unintended outcomes that are essentials on a journey of development. For me, that is a positive view of unwanted outcomes and not the negative view that they are in need of fixing. Unwanted outcomes are OVER and cannot be fixed. Focus on what to do next, not what to fix.

Poor Assumptions

The assumptions I had about the nature of learning for over 20 years were incomplete. It is surprising what can be misunderstood about our natural ability to learn when teachers, coaches, parents, or business training programs are seen as the main source of information and answers for the problem to be solved, without taking the brain's connection to learning into consideration.

Brain-compatible approaches to meaningful learning are emotionally captivating and more a function of student awareness, than direct teacher instruction. Some approaches to learning may offer up little more than a proxy of learning, side-stepping real contact with a student's potential for gaining the kind of wisdom found in flexible knowledge and portable skills (Willingham et al., 2015).

Summary

Some of what I have learned from others includes:

○ Learning is influenced by more than meets the eye; it's an internal process that guides an external outcome after an internal change. What you cannot see going on inside the brain matters and should be taken into consideration when efficient learning, developing, growth and performing are the aim. What we do mentally and physically changes the neuronal connections that hold information inside the brain.

○ I learned that the context and environment, in which approaches to learning and human development take place, are free of frustration and intimidation when brain compatible.

○ From neuroscience, I learned:

The nature of learning is seamless, touching the physical, mental, and emotional components of learning all at the same time. Learning takes place at the intersection of before, during and after acts of learning, teaching, and performing. Learning is based on past experiences.

○ From Dr. Kurt Fischer of Harvard's Graduate School of Education, I learned:

Accurate subject-content information is not the challenge. How approaches to learning are designed and organized to deliver information is the challenge.

○ From Eric Jensen's *Teaching with The Brain in Mind Workshops*, I learned:

The culture and customs of efficient approaches to new learning keep the brain in mind.

○ From cognitive science, I learned:

Our capacity to recall information is much less than our brain's seemingly limitless capacity to encode information. Therefore, it is important that approaches to learning enhance our ability to understand and recall information. I learned that memory is more an act of rebuilding than recalling.

○ From Dr. Robert Bjork, I learned:

What takes place during lessons, wanted or unwanted, is not an indication that learning has or has not taken hold.

○ From developmental research I learned:

Follow the bouncing ball from information in the environment, to the brain, to prior knowledge, to interactions, to outcomes.

○ I have learned:

Brain-compatible approaches to learning support meaningful learning; they do not guarantee learning.

○ From studies into emotions, I learned:

Doing things for their own sake, not a reward, emotionally supports the nature of learning, and to be curious about information. The power of the emotional limbic brain is far reaching and is always a large component of learning.

○ I have learned:

Effectively and poorly managed information and skills start with how the approach to learning interacts with the internal workings of the brain.

Your Notes

CHAPTER TWO

What Influence Does Emotion and Stress Have on Learning and Performing?

It depends!

> **Hebron**
>
> *"When educators fail to appreciate the importance of students' emotions, they fail to appreciate a critical force in student learning. One could argue, in fact, they fail to appreciate the very reasons that students learn at all"*
> —Immordino-Yang, 2016, p. 4.

Hebron

Emotions are central to all of our thoughts and actions. Emotions are reactions to our personal impressions of events by releasing chemicals throughout our central nervous system. These chemicals will then influence our on-going and future responses to any circumstances.

When I spend time with students, the session is designed to be an emotionally safe, positive learning experience. Students are not doing things wrong in my view; they are doing things that are just on a journey of growth and development.

In my past approach to instruction, I would point out what was going wrong, clearly a negative message to students. Today, I ask students to see unintended outcomes for what they actually are, long or short, etc., and not be judged as wrong or poor.

Positive approaches to learning create a space where ideas are not over-explained. These environments are free of the harm that preparation only for success can create. A playful, curious trial-and-error journey is more useful to the nature of learning than keeping an eye on the end result, overlooking the journey.

One of the aims of a positive approach is avoiding the all-or-nothing, right-or-wrong attitude that turns many students into dependent beings lacking in adaptive imagination relevant to life's daily conditions.

Immordino-Yang, "Birds fly, fish swim, humans feel." Emotions become the steering wheel every day for the direction of our future thoughts and actions, including when learning. For example, we typically say we like the book, the movie, or the outcome of our actions, but we actually like the way all these make us feel. Learning should promote positive feelings not negative thoughts.

Yazulla

Dr. Bruce S. McEwen—it is fitting that a chapter on Stress, Learning and Performance starts with an acknowledgement to the neuroscientist who led the way to show how an unrelenting barrage of stress hormones can alter brain chemistry and wiring, with drastic effects on the body, leading to disease, depression, obesity and more. What follows is an edited version of an Obituary in the New York Times (Feb 10, 2020).

> Dr. Bruce S. McEwen died on Jan. 2, 2020, at 81 years of age; he was the Alfred E. Mirsky Professor and Head of the Harold and Margaret Milliken Hatch Laboratory of Neuroendocrinology at Rockefeller University in New York.
>
> It was common medical thinking, dating to the 1910s, that stress was the body's alarm system, switching on only when terrible things happened, often leaving a person with an either-or choice: fight or flight. Dr. McEwen trail blazed a new way of thinking about stress. Beginning in the 1960s, he redefined it as the body's way of constantly monitoring daily challenges and adapting to them.
>
> Dr. McEwen described three forms of stress: **good stress**—a response to an immediate challenge with a burst of energy that focuses the mind; **transient stress**—a response to daily frustrations that resolve quickly; and **chronic stress**—a response to a toxic, unrelenting barrage of challenges that eventually break down the body.
>
> Dr. McEwen's and his research team at Rockefeller University in Manhattan discovered in 1968 that stress hormones had a profound effect on the brain. They demonstrated that toxic stress atrophied neurons in the brain's memory and learning center, near the hippocampus.
>
> Their findings paved the way for later discoveries by other scientists that toxic stress expands neurons near an area of the brain that promotes vigilance toward threats, the amygdala. This area is very important for a subconscious decision to identify a stimulus as threat *vs* non-threat. For example, is the object a real snake or a rubber snake?

Their discoveries ignited a new field of research, one that would reveal how stress hormones and other chemical mediators change the brain, alter behavior and impact health, in some cases accelerating disease. In recent decades Dr. McEwen devoted his energies to understanding how factors like nutrition, physical activity and exposure to early-life trauma can also alter the brain.

"Everything in the field of stress bears his intellectual footprint," Robert Sapolsky, Professor of Biology, Neurology and Neurosurgery at Stanford University said regarding Bruce McEwen.

We cannot overestimate the importance of one's state-of-mind in learning and performing situations. All other things being equal, a person's emotional state at that moment, largely determines the outcome of any action. Emotion is an umbrella that overshadows all other contingencies of motivation, learning and performing.

Hebron

Students not only hear information, but they also have an emotional reaction to the information and how it is presented. Dr. Dave Alred, author of *The Pressure Principle* pointed out "Words are the most powerful performance enhancing drug." Words can emotionally suppress or support learning and performing.

We all have heard, "it's not what you say, it's how it is said that's important." By using the kind of words and strategies that connect with the nature of learning, educators can be more effective in helping students encode, remember, and recall information and skills.

Emotional effects have priority over cognitive processing. Emotions override the influence of the brain's frontal cortex where planning, organizing, and learning take place (Dr. David A. Sauceo). The process of effective learning is free of negative judgments in a safe, smart, playful learning environment.

Yazulla

Just what is a Stressor? People have a gut feeling about stress and its implications, but just what is a stressor? For scientific purposes, a stressor is any condition or stimulus that alters the resting state of equilibrium of the body. All organisms have a resting, basal metabolic rate (BMR), at which all systems are in a state of equilibrium. This is known as *homeostasis*; perhaps as you are sitting comfortably in a chair, just relaxing after a light meal. Simple examples that would disturb this equilibrium include a drastic change in temperature, thirst, perceived threat, pain, sexual opportunity, nausea, stomachache, etc.

When subjected to a stressor, all organisms have mechanisms that are activated to restore equilibrium; for example, shiver, sweat, eat, drink, run, fight, etc. The response to a stressor requires the activation of neuronal, hormonal, and motor systems throughout the brain and body to restore balance. Once balance is resorted, the brain and body return to a resting state of equilibrium, until the next disturbance. What does this have to do with Learning and Performance? As it turns out, a great deal.

"Human movement reflects the emotions of the mind"
—Leonardo da Vinci

It is important to note that we are discussing the effects of everyday types of stress on the brain and body that are encountered in a coaching or academic environment. We are not including the extreme cases of stress that are associated with long-term emotional and physical abuse or substance addictions. There are also memories associated with strong emotional experiences. These have been referred to as 'flashbulb' memories. They can be intense childhood memories, or "where were you when Kennedy was shot?" ''where were you during 9/11?'' Like any other one-trial traumatic memories, we are not discussing them either.

There are numerous books that treat these types of extreme stress. Two come to mind: *The Body Keeps the Score* (von der Kolk, 2014; back on the *NY Times* Best Seller List for 132 weeks as of 3/19/23) and *The Dopamine Nation,* (Lembke, 2021). These books relate case studies of extreme abuse, physical, sexual, psychological and substance. The stories of trauma are heart-rending as the authors describe how the effects of trauma on brain chemistry and nerve networks are serious and often permanent, with drastic effects on personality and behavior. Each book presents chapters on how to deal with the long-lasting effects of extreme stress. With these caveats in mind, we proceed to discuss daily stresses and their effects rather than extreme stress. Still, unrelenting minor daily stresses take their toll.

Recall that a memory involves the reorganization of brain circuits in response to some stimulus. Learning requires the formation of a memory, while performance requires you retrieve then reconstruct elements of the memory, activate motor circuits, and put them into action. All of this activity in the brain occurs as a complex interaction of chemical and electrical signals among nerve cells throughout different regions of the brain.

Any event that alters the electro-chemical interaction of brain nerve cells affects memory formation (i.e., learning), and memory retrieval (i.e., performance), positively or negatively. The most common intervening factor is stress, with its emotional companions, anxiety, and motivation. There is a very large scientific literature on the effects of stress on learning and performance.

Often, learning and performance are treated separately because the effects of stress on learning and performance have some subtle differences. Stress or anxiety occurs when you anticipate a future event that is based on some past experience. It may be good or bad, either your own experience or as observed in others. For simplicity, a stressor can be physical or psychological, positive, or negative, as well as in degree: mild, medium, or severe. This is already a rather complicated 2x2x3 matrix for any experiment to address. Hunger, thirst, fatigue, and fear are among the many stressors that can alter one's sense of well-being and will affect learning and performing.

We use experience to plan a course of action, leading to a Prediction for the outcome. The Prediction sets up an Expectation for some hopefully desired result. Once an action is started, for example a long putt, the anticipation is accompanied by heightened neurochemical and hormonal activity in the brain and throughout the body. If a positive expectation is met, one may feel good; if not, one may feel bad. The feelings are caused by chemical changes in the brain depending on the outcome. These brain changes are automatic, beyond conscious control, and have major effects on subsequent performance. Each person differs in their response to anticipation and anxiety. A consistent failure to meet expectations often leads to frustration and a decrease in the enjoyment of the activity. If you are indifferent and have little or no stake or expectations in the outcome, then there will be minimal emotional response.

Contrast a person who is indifferent to the involved fans at any sporting event in the last seconds of a close game or race. Whether the game is football, baseball, basketball, soccer, golf, etc., once the ball is launched, with win or lose depending on the outcome, involved fans all show increasing anticipation. For example, in basketball, fans in a 1-or 2-point game may be in a near frenzy in the last seconds as a player on the losing side is desperately trying for the one shot that will determine the game.

Once the ball is in the air, fans may take a long, deep breath or simply hold their breath, watching and waiting anxiously for the outcome. The heightened physiological and hormonal responses to anticipation will be about the same for all fans, regardless of the team or athlete they are cheering for. But soon as the ball lands, in or out of the basket, that common emotional anticipation will turn instantly into elation for some and dejection for the others.

The physiological changes from anticipation in the brain and body are quite different for elation joy and dejection sadness. The neurochemical and hormonal responses affect heart rate, blood pressure, digestive processes, mood, and self-esteem. These feelings can last for quite a while and have given rise to celebrations, riots, and domestic abuse. The point is that whether you are a participant or observer, the response to anticipation of an outcome is rather the same for all. However, once the outcome occurs, the behavioral response depends on your expectation and desire. Regardless of whether the outcome is positive or negative, your emotional response will persist for some time and can influence any subsequent decision or action. Overall, response of dejection lasts longer than the response of elation. It takes longer to get over bad news than good news. This is also true for classroom, performing arts, sport learning and performing environments.

> *"In every case, the joy is greater, the more intense the anticipation that preceded it."*
> —St Augustine of Hippo

The same can be said for sadness Why is this? Anticipation of the outcome of events, as described above, is often felt as pleasurable, and is associated with an increase in brain dopamine levels. The increase in heart rate, breathing, and dry mouth is due largely to the fight-or-flight reaction of adrenaline release. This reaction is the same for all of those awaiting the outcome of the event. The reaction to a 'wanted' outcome is accompanied by a further increase in brain dopamine as well as other neurohormones. However, the reaction to the 'unwanted' outcome is accompanied by a reduction in brain dopamine, while the adrenaline release continues. Thus, a dopamine crash, along with maintained adrenaline levels, results in a depressed mood and a maintained fight-or-flight reaction.

The response to the letdown lasts longer than the emotional high. Another example, familiar to most, is a post-holiday slump. The anticipation leading up to an event: Christmas, Thanksgiving, a birthday party, a wedding, a birth is often followed by an emotional downer (dysphoria). Elation can only last for so long as the brain's response to the increased dopamine gets less and less with time. This decreased response to the continued high level of dopamine by the brain is the basis of tolerance; it takes more and more stimulus to achieve the same effect. The response to dejection may be seen as being down in the dumps, with the blues and can last as long until a new positive experience pops one out of their funk. Note: Intense stress affects judgment, which in turn affects decisions and their consequences.

Each person differs in their response to anticipation and anxiety. The same chemicals in the brain often have an opposite effect (i.e., serotonin) depending on which area of the brain is involved with the outcome of the action. The ability for students to devalue their personal stake or reputation in their Prediction and Expectation will mute the hormonal response to Anticipation, and the subsequent hormonal response to an unintended outcome. This attitude will keep stress hormones in check while increasing the release of feel-good hormones. The result will be less stress on the heart, lower blood pressure and blood sugar, leading to more enjoyment and a healthier life experience. What am I going to get out of this lesson? It depends largely on your attitude, what you bring to the instructional table, and how information is shared.

Hebron

Stress produces a positive or negative emotional response. The emotional response depends on the interpretation and expectations of the event which caused the release of stress hormones into the brain and body. To come back full circle, stress may help or hinder learning and performance, which means that your emotional response to any event has consequences for how well you learn or perform. The ability of an instructor or coach to counter heightened expectations with more reasonable ones is part of a Brain-Compatible Learning approach.

In terms of academic, coaching or any learning environment we propose that a strategy, which is critical or negative, normally does not produce optimal learning and performance. There are a few situations in which fear or anger inspire great effort of physical strength, as observed in instances of self-defense or rescue. However, these examples may apply to brute-strength sports, such as weightlifting, but are less likely a factor in most learning and performing situations.

Yazulla

Stress can be thought of as "Arousal," that is the priming of the mind and body to act. In this manner, 'stress' and 'arousal' will be used interchangeably. When one looks at the effects of stress/arousal, there are six timelines when stress may be present:

1. Before the learning period

2. During learning

3. After learning

4. Before the performance

5. During the performance

6. After the performance

Stress and Learning

Before learning, eagerness, and the curiosity to learn new things are the most important positive 'stressors' to increase arousal prior to entering a learning situation. These attitudes of positive anticipation cause the release of 'feel-good chemicals' in the brain that increase alertness, promote learning, and induce a pleasurable feeling. Other factors in this positive attitude are the anticipated quality of the learning environment and competence of the instructor.

During learning, these same stress factors also are at play, but now include the perceived quality of the instruction as well as the attitude of the instructor. Progress and encouragement promote learning, while frustration and criticism interfere with it. The goal of the instructor here is to keep the students engaged and interested. Information presented too slowly or quickly taxes one's attention and promotes daydreaming. Nothing kills a learning situation faster than apathy, whether in the student or instructor.

Enthusiasm by the Instructor in the subject matter is always conveyed to the students who often are drawn along with it. Does that always work? No! At least the student can put the blame on the subject matter on not on the effort expended by the instructor to make it interesting or relevant.

Following learning has its own stress. "Did I get what I came for? Was it worth my time? That was great! I can't wait for the next lesson." Each of these statements expresses an emotional response, which generates a chemical effect in the brain that is experienced as some loss or gain. This attitude following any lesson sets the backdrop for the 'stress' preceding the next lesson and so on, eventually to the performance. Overall, learning is enhanced when positive attitudes and methods of student, instructor, and environmental (physical and social) are combined.

> *"You have power over your mind—not outside events.*
> *Realize this and you will find strength."*
> —Marcus Aurelius

Stress and Performance

The effect of stress on performance is similar to but a bit more involved than for stress on learning. Recall that learning and performance are directly related to one's motivation to do so. Before a performance, one's motivation is reflected by a level of arousal to act. There is a great deal of research showing that the effect of stress depends on how difficult the task is or perceived to be. The earliest expression of this relationship was the Yerkes-Dodson Law (Yerkes & Dodson, 1908; Corbet, 2015, for review); it stated that performance on any task initially improves with stress, but high levels of stress impair performance on a *difficult*, but not on an *easy* task. For simple, non-complex tasks, or those that are very well learned and rehearsed, performance improves as arousal increases, and continually increases to level off even with high levels of stress.

Presumably, a high level of confidence and skill allow the behavior to kick into automatic. This differs for detailed, complex, or poorly rehearsed tasks for which performance initially improves with arousal and then decreases as arousal/stress becomes high. Consider threading twenty sewing needles by hand. The task would get more and more difficult under an increasing time pressure, from say, two per min to ten per min. There is more pressure to thread twenty needles at six seconds each than thirty seconds each, even more so, depending on the amount of reward or punishment for success or failure. The basis for this dose-dependent effect of arousal on learning or performance is the short-term release of stress hormones, particularly adrenaline, which initially primes the body for action, fight-or-flight response.

However, heightened, or persistent stress causes the extra release of cortisol, which interferes with the formation of memory (learning) and retrieval of memory (performance) of behaviors that are not related to the stressor. The main effect of the heightened stress is distraction, shifting attention to survival, that is, the payoff or punishment, rather than the task at hand. It is important to note that multiple areas of the brain are involved in the transition from mild to heightened stress, as in the hippocampus and amygdala, also including the hypothalamus, pituitary gland and adrenal glands, the HPA axis (Diamond et al., 2007, for extensive review).

Stress following a performance is usually due to its evaluation either by yourself or others. The response may be satisfaction or disappointment, which may or may not reflect the outcome of the performance. One's attitude can crush one's confidence, spur one to greater effort to improve, or enhance one's confidence. Each of these responses has effects in the brain and will color the manner in which the next performance is approached.

Learn to Perform Under Stress?

So far, we have been treating Learning and Performance environments as separate and sequential: performance follows learning. An intriguing question was posed. Since any performance involves some level of stress, how does one learn to perform under stress? Should the learning stage include the kind of stresses to be encountered during a performance? If yes, then when? Or should one rely on experiences during performances to learn how to cope with the accompanying stresses?

We already know from the Yerkes-Dodson Law, and our own experience, that all learning is enhanced with mild stress. The stress focuses attention and stimulates retention. However, at high levels of stress, attention shifts from learning to self-preservation, and the learning of complex tasks is suppressed. So how do you learn to perform under stress?

In academia, all other environmental factors being equal, learning depends on the motivation of the student to learn the material. The stress can be increased by imposing a time limit to learn the lesson content. This does not stimulate learning; it only selects for those who learn more quickly, regardless of their motivation. The same is true for imposing a reward or punishment. For academia, learning will not be enhanced by simulating very high stressful situations to be encountered in any kind of evaluation procedure. However, once the material is learned, you can simulate the stress of performance in practice sessions.

These types of simulations are common in courses and practice exams given for qualifying exams for college (SAT), graduate school (GRE), medical school (MCAT), law school (LSAT), certified public accountant (CPA), and so on. These practice exams simulate the content and time constraint of the exam, but they lack the win-or-lose payoff for the actual exam. Similarly, in public speaking, oral exams, stand-up-comedians, political speeches, and so on, practicing learned material for the performance is learning to perform under stress, but it is not learning under stress. Thus, no learning environment should increase stress beyond mild, to the point that learning is suppressed. It does not work; it is counterproductive.

I expect that this would apply to any complex motor activity, sports, trades, arts, medicine, sciences, etc. You cannot terrorize someone into learning how to play any musical instrument with a level of competence and enjoyment. That does not teach them how to perform under the stress of a public performance. Only confidence in their ability and experience will teach them that. In sports, you cannot simulate the stress of a million-dollar golf shot, or field goal, or penalty shot, in a learning situation. We often hear of performers in the arts suffering jitters or butterflies, to the point of nausea, and even vomiting, prior to a performance. It is only with experience of overcoming stress in real-life situations that one truly learns to perform under stress.

A good example of this is a Championship in any sport: tennis, golf, baseball, soccer, hockey, basketball, etc. Sport announcers routinely make a distinction between rookie first timers and veterans in the big games. The reason, of course, is that the veterans are expected to handle the stress of big games better than the rookies because of their experience. The veterans are expected to be a calming influence on the rookies. The choice of the captain as in the Ryder Cup is particularly important as a stabilizing influence, as is true for veteran players in the locker room. Team Captains are chosen just for this purpose.

The attitude of leaving the past behind is perhaps best illustrated by Mariano Rivera, the premier closer for the New York Yankees. Despite all his saves, I could not forget his blown save in the 9th inning of game 7 in the World Series against the Arizona Diamondbacks. Although I remembered it, Rivera was able to move on as illustrated in an excerpt from a *Wikipedia* entry:

> Mariano Rivera exhibited a reserved demeanor on the field that contrasted with the emotional, demonstrative temperament of many of his peers. Hall of Fame closer Goose Gossage said that Rivera's composure under stress gave him the appearance of having ice water in his veins. Commenting on his ability to remain focused in pressure situations, Rivera said, "When you start thinking, a lot of things will happen . . .

If you don't control your emotions, your emotions will control your acts, and that's not good." His ability to compartmentalize his successes and failures impressed fellow reliever Joba Chamberlain, who said, "He's won and lost some of the biggest games in the history of baseball, and he's no worse for the wear when he gives up a home run." Rivera explained the need to quickly forget bad performances, saying, "the game that you're going to play tomorrow is not going to be the same game that you just played." Derek Jeter called him the "most mentally tough" teammate with whom he had ever played. When asked to describe his job, Rivera once put it simply, "I get the ball, I throw the ball, and then I take a shower."

We cannot overestimate the importance of one's state-of-mind in learning and performing situations. All other things being equal, a person's emotional state at that moment, largely determines the outcome of any action. Emotion is an umbrella that overshadows all other contingencies of motivation, learning and performing.

Hebron

Suppressing Learning

Perhaps the most efficient way to support learning is to have insight into what will suppress the experience of learning, memory, developing and growth. What follows is influenced by *The Handbook of Neuroleadership* (Rock & Ringleb, 2013).

The way humans relate to each other and themselves influences what we can and cannot do physically and mentally at the emotional-biological level. Any provider of information during a learning opportunity is dealing with a "social brain," which is also an "emotional brain." Learning and emotions cannot be separated.

NOTE: The organizing principle of the brain: minimize threat—maximize reward.

When I first read that statement, insights came to mind that were game-changing for me and the students I was spending time with. The social needs of students should be treated in much the same way as the need for food and water, says Dr. Rock. He pointed out that knowing the drivers of what can cause a threat response when learning, enables people to design interactions to minimize that outcome.

By knowing how these drivers draw conscious attention to an otherwise sub-conscious process is knowledge of what supports learning effective skills, minimizing frustration, intimidation, and maximizing reward responses. For example, during learning, a lack of student autonomy activates a genuine THREAT response. Leaders and educators may want to consciously avoid micromanaging information given to employees or students, thereby giving them more self-learning opportunities. Knowing about the drivers that activate a reward response enables teachers to motivate students effectively by tapping into internal rewards, thereby reducing the reliance on external rewards and external information.

NOTE: our limbic, emotional system processes incoming stimuli before it reaches conscious awareness. It is an alert system; something has happened; is it friend or foe. Brainstem limbic networks process threat and reward information within a fifth of the second, providing ongoing unconscious information of what is influencing you in every situation of your daily life.

Summary

Stress is a consequence of us being alive and reacting to changing environmental cues. The motivation to get up and move is due to the stress induced by some situation that gets our attention to be acted upon. Under normal conditions, stress is beneficial to maintain physical and psychological balance. Stress is beneficial for learning and performing, but at the right amount and right time. Your level of stress depends on your experience and expectation of the outcome of any event. You cannot control your past, but you can deal with your expectations, keeping it realistic. In this way, the harmful effects of too much stress may be moderated.

We suggest a coaching approach that promotes a reduction in stress by having students learn to accept realistic expectations is one aim. Students should accept any game for what it is, a game, without investing their sense of self-worth in the outcome. Students and instructors should be aware that emotional responses affect brain activity in a way that is very difficult to control, once started.

There are activities with a payoff (success or failure) that many people engage in for the sheer joy of the activity, without much attention given to the payoff, for example recreational fishing. The ultimate goal then for the coach is to instill in recreational golfers' enjoyment of the game of golf for itself; the outcome of the score may improve but becomes secondary.

"You cannot change what's over, only where you go"
—Enya, Pilgrim

Your Notes

CHAPTER THREE

How is Learning Determined?

It Depends

> *"Although the learning-performance distinction is overwhelmingly supported by empirical evidence, there appears to be a lack of understanding on the part of instructors and learners that performing well when training at times does not support learning."*
> —Dr. Robert Bjork, UCLA Learning and Forgetting Lab

Hebron—*The Self and Learning*

Let's look at THE SELF as the sum of our emotions, thoughts, and actions. From this view the mind/brain is responsible for coordinating all the activities of the self, including learning and memory. Note that all three, the mind, the brain and the self are influenced by emotions.

The architecture and development of the human mind/brain is a fluid process of self-wiring and rewiring ongoing over time, a natural reaction to our wanted and unintended outcomes in ever-changing environments. Our it-depends, self-development reactions started on the day our first ancestors walked the earth and continues from the day when each individual is conceived and starts to learn.

The brain is the most essential organ when it comes to learning and self-perception. Many body parts could be replaced, and we would still be the same person. But if it were possible to transplant a mind/brain, we would no longer be the same person.

Who we are and what we know depend on how our brain self-wired its information highway in the past and continues its self-wiring from moment to moment. This involves a process that changes the brain connections that store information. The history of our thoughts, self-talk, actions, and the outcomes they produced and encoded in our subconscious minds are constantly changing the brain wiring. Instructional methods leverage that process when they are brain compatible.

The brain has many duties, but first and foremost, it is an information processing organ, evaluating information coming in from the internal workings of our body and our interactions with the external world. Some of what comes into the brain is turned away and some is sent along in the form of patterns, sequences and meaning for future reference during learning. This information is in the form of chemical-electrical messages that are often incomplete. The brain then guesses and fills in, based on prior knowledge.

Everything is the story we tell ourselves. We say this is hard or easy, good, or bad, I like, or I don't like. It depends on the story we tell ourselves, which is influenced by what's going on subconsciously in our mind.

The very nature of the human mind/brain reveals that the eyes do not see, the ears do not hear, the mouth does not taste, we are not catching the ball, swinging the club, or making decisions. It is the mind-brain that organizes and guides all that, including learning.

Two Components of Learning

While the aim of instruction is learning, Dr. Bjork of UCLA Learning Lab points out that how we learn is different than how we think we learn. It helps to recognize that knowing information and knowing how students learn information are two components of the same topic: a) providing instruction, and b) learning and memory of instruction.

The topic of memory is an interesting one and at the core of learning and performing in sports and life. The study of memory is one of the central pursuits of neuroscience. Some of the following insights are based on the book *Your Brain* (by Matthew MacDonald, 2008).

Perhaps one of the elements of human inconsistency is our memory.

It is fair to say our memory is playing golf and doing other things in life, like driving a car. While playing golf and swinging the club, we are led consciously and subconsciously by recalling feels, visualizations, and information.

> *"Memories of experiences and prior knowledge are stored in our brain. When recalling, the brain needs to reassemble information from a vast network of different stored concepts and details. Here is the problem. When we attempt to pull a memory back together, what you get is not the original. In fact, data from Neuroscience points out that remembering is an act of creative re-imagination. That means the recalled memory doesn't just have a few holes and out-of-place pieces, it also has some entirely new bits pasted in memory."*
> —(MacDonald, 2008, p.101).

Memories are never exactly the same, nor are golf shot outcomes and swings, and both have human memory at their core.

This continuous process of brain reorganization underlies all long-term memory, learning and performing. In my view, at times some approaches to instruction overreact to unintended outcomes especially after several wanted results. The unintended outcomes are the result of being human and they will always exist, no fixing needed, just accept, and move on. Why did you hit that unintended shot? Because I am a HUMAN being, not a machine.

When information is first shared with or gathered by us, it enters our short-term memory and is held there for a short time, less than a minute, before it may or may not move on to long-term memory. The information that moves on is not in details, only pointers. What I call general just-in-the-ballpark concepts.

The brain receives patterns of information from the outside world, then develops memories and makes predictions based on mixing what it has experienced before with what is happening now. Confusion or a lack of learning occurs when new information cannot be associated with something similar from the past. This is why the use of metaphors and stories during instruction support deep learning.

Our thoughts, perceptions (the stories we tell ourselves), emotions, and actions are first generated by groups of cells firing together in the brain, leaving behind in long-term memory general patterns made by cell connections. Most of the information coming into the brain and the central nervous system is not retained in long-term learning. Only key general elements are selected to be integrated with prior experiences and knowledge, causing cells to fire together.

SUGGESTION, provide information in ways that it can be added to what students already know.

SUGGESTION, provide instruction that moves beyond just offering information and guides students in the direction of developing and using their own perceptions and feelings to construct their own insights.

SUGGESTION, limit the variety of information given.

SUGGESTION, limit the amount of information given. One way to do this is by "Chunking," a process by which related pieces of information are combined into one suggestion or one picture. Instead of iterating all the steps of the golf swing (stance, head, shoulders, arms, grip, etc.), visualize "just swinging the weight of the club." All the elements of the swing are then grouped (chunked) into one movement.

Memory—Learning to Perform

Yazulla

A Memory may be loosely defined as an alteration in a neuronal circuit (i.e., brain connections) in response to time-related events in paired-environmental stimuli. Depending on the brain chemistry involved, the memory may only last for a short-term, longer-term, or a lifetime. When a memory of an event is encoded in the brain, we normally say that Learning has occurred. But, in order to demonstrate that learning has occurred, the memory must be retrieved from the brain circuitry and expressed behaviorally as effective Performance.

It is important to know that memories are distributed in many related areas in the brain. The size, color, shape, and position of an object is represented in different areas of the occipital cortex, as are sound and motion in their respective areas of the cortex. As a result, a memory is reconstructed by recruiting neural elements from multiple brain regions. A memory is not stored in a single brain neuron. We learn and remember best when we employ multiple modalities, sensory and motor. Remembering (recall) is a consequence of the reconstruction and integration of patterns, sequences, and relationships within the neuronal network activity throughout the brain.

> *"It is beyond doubt that all knowledge begins with experience"*
> —Immanuel Kant

A very simple example of learning is touching a hot stove for the first time; a heat stimulus leads to a rapid withdrawal reflex then the pain sensation hits and bingo, you have one-trial learning. You have learned that a stove can be painful, and not to touch the stove without first checking it. In the absence of recording brain activity, how would I know that you have learned something? I would have to observe your behavior, i.e., a performance when you are faced with a new opportunity to touch the stove again. You may know that you have learned your lesson, but an observer can only determine learning by observing a performance of how you approach and touch the stove. Are you cautious or not? Anyone who is cautious has learned something about self-preservation.

Learning is inferred by observing Performance. The simplest learning is approach/avoidance that is a survival skill for any organism, as in the hot-stove example. Regarding more complex learning, memorization is at the heart of it, whether you are trying to learn directions, a poem, playing a piece of music, a golf swing, a surgical technique, etc. Let us use memorization of a poem as an example. There are four strategies one can use: a) silently read it, b) vocalize it, c) write it down, or d) just listen to it. Once you are satisfied that you have memorized it, you could test yourself by silently reciting it, vocalizing it or writing it down. Each of these strategies requires a different set of mental, sensory and motor skills, the competence of which may differ for any particular person.

Now consider that the person in the above example must demonstrate whether they learned the poem in front of a group by a vocal performance. Standing there silently reciting the poem to themself is not going to work. They are required to recite it, out loud. Two factors come into play here, the method the person used to memorize the poem in the first place and their ability to recite it in front of a group. If the person has not heard their own voice reciting the poem, they could be at a disadvantage.

Alternatively, the person could have stage fright and regardless of how well the poem was learned, performance would be affected by the emotional state. Here is when the instructor/teacher must differentiate the importance of learning in the context of the performance environment. Much depends on what the goal of the learning experience was. If the goal was just learning the poem, then reciting or writing it down would suffice. If the goal was to perform on stage, then the stage fright effort was inadequate. Learning may have occurred, but the performance fell short. This difference sets up the classical excuse of any student *"But I know that!"* Yes, perhaps you do, but your performance did not demonstrate that learning had occurred.

> *"Nothing is as well learned as that which is self-discovered."*
> —Socrates

Hebron

This sounds like golfers who are hitting balls on the range or during instruction with great gusto and satisfaction, then cannot repeat that when on the golf course. For me, this shows learning was not deep. I think learning is defined by performing well. I am not sure we can say something has been learned if it cannot be performed.

When learning is deep, the performance can be wide, flexible, and portable. Deep learning or knowing a lot about one area of a topic can support learning more effectively than knowing a little about several areas of the topic. Deep learning moves beyond just knowing information into an effective application of that information in ever-changing environments.

The distinction between Learning (memory formation) and Performance (memory retrieval) is extremely important to keep in mind because the ultimate goal of most learning situations is to optimize recall and performance. The factors that affect learning are equally important for performance, particularly regarding Motivation, Expectation and Stress.

Information exists in neurons, and at times millions of them connect to remember or perform. We all can learn, and the brain remembers with the aid of Metaphors and Stories:

- New Information must connect with prior information

- When we remember we do not recall we reconstruct

- None of our memory is stored in one single neuron

- Common threads among different kinds of information support memory

- We learn and remember best through multiple modalities

- To be remembered, information needs to make sense and be understood

- Research points out that a student's prior knowledge is essential to help them experience new learning

The brain develops stronger and extended memory circuits when new learning can be connected to multiple circuits of prior information (vision, hearing, movement, or sound).

One example; when coaching sports, common information could be motion and pointing out to students some similar motions in different sports.

When new information is shared and joins an existing pattern that's in the brain, memory networks incorporate it more efficiently.

I like to tell students that there is actually no new learning. When new information comes into the brain it must mix with prior knowledge before it can become learned and remembered.

Have you ever asked something to be repeated because you did not understand what was said? That happens because what was said did not connect to similar information that is understood. Some detailed instructions have this negative result.

Storage of memory in neural networks is based on the brain's recognition of patterns and relationships and sequences, not details. For me this was a huge insight.

Performance Evaluation

"Fear of negative evaluation can take the fun out of anything."
—Anonymous

Yazulla

In one way or another everyone is evaluated, either immediately or in the near future, by ourselves or by others. Humans are social creatures and as such, are constantly scrutinized for all sorts of things: hair style, manners, clothing, occupation, politics, friends, body size and shape, and on and on. The manner of evaluation can have major effects on one's self-image (positively or negatively), future performance, and certainly on the motivation to pursue any particular activity.

Evaluation in academia proceeds in a rather straight line: a) exam grade, b) course grade, c) cumulative grade, d) letters of recommendation, e) next stage, f) continue with evaluations through employment to retirement. Evaluation in Laboratory courses involves a demonstration of mastery in some lab exercises (i.e., biology dissection, chemistry or physics exercise, engineering or computer project, etc.) as well as the content material involved. Laboratory courses are often both content and hands-on driven, in many ways like sports, trade, or music instruction.

Learning in a content-driven course is evaluated by an oral exam or more commonly by two types of written exams: Speed and Power. These can take the form of multiple choice, fill-in-the-blanks, short answers, a paragraph, or an essay. A Speed test is time-limited, and no one is expected to finish it within the time allotted. College entrance exams are like this. Rapid thinking and decision making are rewarded. A Speed test can be very stressful when you realize you have no chance of finishing the test, especially if you spent too much time on earlier questions and cannot get to questions you likely know. The strategy is to answer the questions you know for sure and then go back to the harder questions. This maximizes the number of correct answers while minimizing errors.

In a Power test, enough time is allowed for all students to finish the exam, with even some time to reflect on the answers. As there are fewer questions, the strategy is to read all the questions first, then answer the ones you know to build confidence. Often that will trigger memories for you to answer the more difficult questions. Power tests are often graded more harshly because the time-limit stress has been reduced and sufficient time has been allotted for you to demonstrate your learning.

Problems arise because each type of exam determines learning by using a different measure of memory Performance. Multiple-choice exams test for recognition because the answer is already visible, and the student uses cues to choose which option is correct. Fill-in-the-blank requires recall of words to fit in the context of an existing sentence. Short answers, paragraphs and essays require complete recall in context of the question without any information regarding the answer. The issues are that students differ markedly in their ability to deal with each of these exam types. The excuse is that the student is a terrible test taker.

Out of frustration, sometimes the student simply shuts-down. More recently, such students have been given extra time and or a quiet place to take the exam. Other students do poorly on multiple-choice tests because they find the questions confusing and misleading. These students prefer to take written exams in which they can express their knowledge in a free-wheeling manner. All of these stress-related conditions affect the efficiency with which memories can be retrieved and expressed in the exam environment. These same conditions apply to oral exams, which require complete recall with the additional stress that oral exams are taken in front of an 'audience.'

> *"Positive feedback charges up a worker, negative feedback saps the job of some of its intrinsic motivation."*
> —Unknown

Hebron

Similar to academia, an athletic or artistic performance can be evaluated immediately or in the near future. A major difference is that an athletic performance is physical and mental, and usually in public, not private. For example, in a golf tournament, an evaluation of a golf stroke can be seen by the reaction of the caddie, the opponent, the crowd, the announcer and everyone watching at home. The reaction of the player to these evaluations can set up the emotional response that carries over to the mind set for the next shot. Thoughts between shots could be about placement in the tournament (financial) or just World Golf Rankings (pride). This is true for any sport or artistic performance. Evaluation or assessment of a performance is public, immediate, and often unforgiving. This does not always have to be the case for the player. Use the mindset of acceptance without judgments and move on to the next shot.

Assessment during sports learning should be quite different from a performance assessment; often it is not. Coaching should involve positive-assessment practices, not ridicule, judgment or criticism. Positive-assessment practices engage the athlete in the interpretation and implications of their outcomes when learning. Positive assessment promotes the idea that guided development of the skill (not fixing) is the aim of a student-centered, brain-based learning process. Telling the student what is wrong takes the student's self-assessment out of the learning session. Feedback in itself may not promote learning unless students are positively engaged with it. In *The Learning Coach*, (Jones et al. 2014) point out that regardless of the demand on the coach, positive-assessment learning is central to quality coaching, even at the high-performance end of the spectrum.

Learning and Pedagogy concepts traditionally have been outside the domain of sports coaching. However, more contemporary research has demonstrated that coaching is really about learning. Pedagogy relates primarily to the interactions between the coach and player. A Learning message refers to the information (explicit and implicit) necessary for developing performance, including: the skills, tactics, and rules. Failing to recognize this key aspect of learning and pedagogy overlooks opportunities for optimizing coaching and athlete development.

University of Queensland

The following is based on research from the School of Human Movement Studies, University of Queensland, St. Lucia, Australia. The positive-assessment approach can support the learning of:

1. targeted knowledge

2. process

3. skills

Positive Assessments support having students assess and apply the necessary *what to do* knowledge during learning for themselves, not *what to fix*.

The term "Feedforward" was coined by Peter W. Dowrick in his Ph.D. thesis at the University of Auckland to indicate the notion of using positive-assessment feedback to assist in the improvement of future performance, allowing athletes to become assessors of their own learning.

Feedforward provides information, images, etc. about future possibilities, rather than focusing on past performance as in the case with feedback. Through this positive process of guided self-reflection, athletes learn to access, interpret, and use positive feedback about what to do, not what to fix to improve future performances. Dowrick, (1999) has argued that an absence of an investment in the understanding and refining of positive-assessment practices is a notable oversight in the field of sports. This significance is a valuable condition for ensuring that any assessment decisions are learning oriented.

Positive approaches to learning humanize a journey of change. Positive-learning opportunities are engaging and emotionally safe. They include endless trial-and-error outcomes that have supported human development from infancy to adulthood and into old age. In positive-learning environments, there is nothing to fix; unintended outcomes are seen as part of a journey to full development and lead to deep learning of what to do differently.

This approach to teaching frees teachers from acting like puppet masters, who are fixated mainly on the future and overlook the real value of a curious, playful journey in the direction of full potential, yet to be experienced.

In environments that offer deep learning opportunities, teachers and students perform together and grow separately. Any decisions made are dictated more by circumstances than ideas about end results, which unfortunately, can create limits of growth and enforce conditions that do emotional harm.

Yazulla

Positive assessment in academia often takes the form of quizzes, short tests on recent material presented in the course. Quizzes provide immediate feedback on a student's grasp of the material, without having a major effect on the final course grade. Ideally, positive assessment, without consequence, takes place in the exchange between faculty and students during the class and scheduled review sessions.

A major effort in schools and other training environments is being directed at improving the conditions in which learning occurs in order to increase the efficiency of learning. The obvious question then is how do you objectively determine whether learning has taken place, and if so, how much and how long will it be retained? To this end, students and teachers in some school systems can be obsessed over content in common core curricula, scores on standardized tests, graduation rates, admissions to prestigious schools, etc. This has led to the practice of "Teaching to The Test" by which the curriculum/syllabus is designed to meet a standardized test, whether the contents of the test are in line with the goals of the teachers. The Test then determines the curriculum.

Entire Departments of Assessment have been set up to measure student learning and, by inference, faculty performance. Learning cannot be measured directly. Rather it is inferred by observing PERFORMANCE. Performance engages the motor system involved in speech, running, hand movements, etc. This means that regardless of how new information has been encoded and stored in brain circuits as memory, performance depends largely on the ability to retrieve and reconstruct memory elements, and then engage the neural circuits used to express that learning. The choice of test format plays a large role in the performance of many students, often to their disadvantage.

> *"Good judgment comes from experience and a lot of that comes from bad judgment"*
> —Will Rogers

Hebron

I found it is effective for students to be guided in the direction of developing the mental tools to accept the outcome of their own actions without judgments. We propose that in any instructional or performing environment, negative reinforcement judgments, criticisms, etc. are counter-productive to success and self-esteem.

Suggestion: Again, the unintended outcome can be the lesson for what to do differently.

Students are quite capable of beating themselves down when their learning and performance are not up to a standard, without any additional emotional abuse of the instructor, parent, or peer. We state that it is beyond debate that positive reinforcement is more effective for learning than negative reinforcement. Yes, we know there are one-trial, horrific learning situations that persist for a lifetime, referred to as "flashbulb memories." But we are not referring to post-traumatic-stress-syndrome (PTSD).

Yazulla

"Don't touch that." Negative reinforcement is more useful in a pain-avoidance situation. It is more efficient to learn by the negative experience that a stove is hot than to be rewarded for not touching it. You would not be quite sure why you should not have touched it unless you actually found it hot by touching it. In these cases, positive reinforcement (reward) is not the easiest way to foster learning. Positive reinforcement here to *not* do something is often unlikely to overcome the curiosity of "what does this do?"

Thoughts on Curiosity

"This experience sufficiently illuminates the truth that free curiosity
has greater power to stimulate learning than rigorous coercion"
—St. Augustine of Hippo

Hebron

Every month *The Wall Street Journal Magazine* features well known individuals to give their view on a topic. The topic for the Aug 1, 2016, issue was 'Curiosity'.

Actor Joseph Gordon Levitt said, "Problems arise when you get too attached to an answer, even in the face of new information or experiences. I find that among the people I know, the most intelligent ones are those who ask questions more than they make statements. They remain curious, even about things they know to be true."

Artist Taryn Simon said, "I was pushed to look beyond the visible, everything had an under-belly. . . . But humans crave certainty, no matter how falsely based."

Jane Goodall said, "Intelligence is the way you express curiosity and the lengths you go to satisfy it. We're always coming up with new questions."

Author Mark Haddon said, "I love the idea that curiosity isn't always about increasing your range or traveling to new places. What is already there can be enough to be curious about—being open to the world that is around you."

Yazulla

Curiosity is the urge to explore objects or places that are novel. What is over there? What is that? Why did that happen? What if . . . ? All progress, be it agricultural, technological, scientific, medical, athletic, or social, is the result of someone satisfying their curiosity by seeking the answer to some question. Or, as posited by George Bernard Shaw below, the same can be said for one who is discontented and wants to change their environment, make things easier or more pleasant.

> *"The reasonable man adapts himself to the world: the unreasonable one persists in trying to adapt the world to him. Therefore, all progress depends on the unreasonable man."*
> —George Bernard Shaw

Any animal, humans included, will stagnate in a constant environment because there is nothing to be curious about. A particularly sad example was Gus, a polar bear (1985–2003) in the Central Park Zoo. After two female companions died, Gus started unusual behavior. I remember seeing Gus just swimming back and forth in his small enclosure, over and over and over again, just back and forth. Gus did this 12 hours a day, 7 days a week. I remember people enjoying this because the bear was actually doing something rather than just sitting or sleeping on a rock. Gus became a celebrity and ticket sales soared. I just felt sad watching all of this. Gus's obsessive-compulsive behavior was viewed by the press as symbolic of the stresses living in New York City. Gus was bored, deprived of any opportunity for curiosity and expressed his depression by ceaselessly swimming.

Efforts made to enrich his environment improved his behavior somewhat but did not eliminate the obsessive swimming. Gus was diagnosed with an inoperable thyroid tumor and euthanized. There are numerous examples of similar behaviors in confined zoo animals. The same applies to people that are isolated for extended periods of time. Sensory deprivation is one form of torture based on such a need for variety.

The brain demands change, a variety of sensory stimuli for proper development and health. The seminal work in the 1960s and 1970s from the laboratories of Harry Harlow, David Hubel, Vernon Mountcastle, Roger Sperry, and Torsten Wiesel showed the causes and drastic effects of sensory deprivation and emotional deprivation on brain development, function, and behavior. These studies explained the causes of prolonged (several years) extreme sensory neglect on the reduced brain size and mental development of infants, effects that lasted into adulthood. These studies led to the appreciation and introduction of "enriched environments" and emotional support for babies and children in hospital nurseries, homes, and schools.

Satisfaction of curiosity always involves an unknown outcome. For every action you engage in, there is some conscious and unconscious Expectation as to the outcome. A reason for this is that we live in an orderly universe in which there is clearly defined cause and effect notwithstanding

Chaos Theory, (but that is another story). Time moves forward not backward. Humor and the appeal of magicians depend on misdirected expectations. Randomness and miracles upset the balance of cause and effect. A shattered bottle of milk does not reconstitute itself. Our entire life from day to day depends on acts of faith that events will follow in a simple cause and effect. The sun rises; it sets. We let go of something, it will fall. In addition to the orderly progress of natural events, we proceed through the day by playing the odds that violations of socially agreed behaviors will not occur. Despite uncertainties, curiosity involves an expectation that the outcome will fall within the bounds of cause and effect.

> *"Chance favors the prepared mind."*
> —Louis Pasteur

Hebron

Any Expectation inevitably leads to the Hope that something will or will not occur. For example, you strike a golf ball; it will move. An Expectation, based on your experience, is that the ball will go somewhere in your intended direction. You do not know this, but the Hope is that it will. Ball flight in golf cannot be controlled. If it could, professionals would not miss greens, fairways, and putts.

Suggestion: Before we swing, a player can pick a club and plan what they want to do. But once we address the ball, just allow the swing to take place and accept the outcome. There is always some uncertainty as to the outcome. Contemplation of this uncertainty leads to Stress. Depending on the outcome, your feelings may be positive or negative. Your emotional response to the outcome then becomes part of the experience that will impact your expectation for the next golf shot. Suggestion: Just accept the outcome and move on to the next swing; no judgments, no excuses, no explanations.

Yazulla

The same is true in an academic situation. You study hard for an exam and expect to do well. However, you do not know exactly what questions will be asked or whether they will conform to what you studied. Your performance depends on how well the questions are in line with what you have prepared for. The point is that your actions do not occur in a vacuum. Rather, each decision and action sit upon the sum total of your experiences, in context of the environment (physical) and social pressures (mental) in which they occur. High expectations lead to emotional highs and lows depending on the outcome of your actions. Muting your expectations should reduce the emotional response to the outcome; and this in turn will keep you focused on the task at hand. Outcomes are what they are; they are not to be judged but accepted.

Hebron

I found the following on GetSportsIQ.com. "Learning is more efficient when using a plan for learning than a course outline, and also when guided by more of a process structure than content structure." (Knowles, 1983).

The technical knowledge of the skills fills volumes of books, hundreds of DVDs, and dozens of training seminars and conferences. But that is only part of the picture. Just because a person has great knowledge of the sport, does not mean they are the kind of effective coach who:

○ Cherishes the person over the athlete.

○ Knows that being an athlete or student in school is just a small part of being a human being.

○ Never does anything to advance the athlete at the risk of the person.

○ They don't tune out an athlete's worries, fears or mentions of injury.

○ They understand the importance of emotional connection. When we feel safe, we can trust and when we trust we can learn.

○ They know that this foundation of trust is essential. They connect before they direct.

○ They start with the end in mind and keep their attention on the big picture and the goal of the athlete.

○ They have a plan, but are flexible, as they are aware that the road to success is filled with twists and turns.

○ They support athletes' struggle. They understand that learning is a curve. Like a muscle needs to break down before building up, students need struggles to push forward. Effective coaches and their students do not panic when this struggle happens.

○ Effective coaches state what is positive. They say what to do, rather than "don't do this." "Don't bend your arms "is not effective feedback.

○ They find the bright spots and build from there and are aware of what needs development. They try to improve standards but put more attention on the areas that an athlete excels in.

○ They do not try to break bad habits; rather, build new habits.

○ They know that the most effective way to develop a workable motion is to train what works, and not on what is poor.

○ These coaches do not use scores or win-loss records as their sole measure of success. They understand that doing so can erode the long-term development of the athlete. Instead, they develop competencies for the long run, even if that means sacrificing success at the beginning of the journey. *If you had to choose, would you rather have your child be the strongest student in the first grade or in the twelfth grade?*

○ They use the right mixture of attainable and desirable goals. These coaches have zoned in on the sweet spot of and value challenge.

○ Efficient coaches appreciate natural aptitude but know that this can only take an athlete so far.

○ Effective coaches concede that coaching extraordinary talent is not a fair assessment of their value as a coach; rather, they measure their coaching efficacy by spending time with an athlete who is less gifted and helping that athlete succeed. How many beginners that we coached are still playing years later may answer how effective a coach we are.

○ Effective coaches are always trying to figure out what makes their student tick so they can better reach them.

○ Separate training and learning from practice. Effective coaches understand that practice begins after training and the athletes have learned. Train what you are going to put into practice. Use the mindset that training is acquiring, practice is applying.

○ Effective coaches support human development. They have a working knowledge of the milestones of human development and tailor their actions and expectations to meet the athletes where they are.

○ They use positive coaching techniques and do not yell, belittle, threaten, or intimidate. They do not need to bully to get results. While short-term success may occur under such pressure-filled environments, an effective coach knows that in the long run these techniques will backfire and are dangerous to the development of the student.

○ They have a learning, developing and growth mindset. These coaches believe that our basic skills can be developed and grow over time. They reinforce this with their athletes over and over, so their athletes feel motivated and are productive. "You can do this," is their message.

○ Effective coaches are humble enough to admit that they are not perfect.

○ Effective coaches go beyond instructing their athletes; instead, they educate them in skill and age-appropriate ways regarding the purpose of their objective.

○ They understand that interest and fun are essential elements in training, no matter how elite an athlete becomes. The number one reason that athletes quit sports, even sports that they love and in which they are succeeding, is because it is no longer fun. FACT!

○ Interest and fun are not frivolous: they are the foundation of an athlete's healthy commitment to a sport. An important insight, fun follows an interest; keep individuals interested during unintended outcomes and they will have fun.

In the 21st century it is now accepted that content knowledge is just the beginning of what makes an effective coach or teacher. Yet absent other coaching qualities, namely insights about learning and memory, just content knowledge does not support effective coaching.

What's referred to as the "biology of learning" is providing rich insights into how we can have a better model and effectively create long term memory as part of a learning experience. One new model is referred to as AGES.

The **AGES** model is a four-part approach for making learning stick by activating the brain's hippocampus.

1. **A**ttention

2. **G**eneration

3. **E**motion

4. **S**pacing

Attention. Receivers of information need to be paying attention for the hippocampus to be activated sufficiently for topics to be learned.

When students are not paying attention, neurons decrease their firing and learning decreases significantly (Kensinger, Clark, & Corkin, 2003). This has been shown to occur even from small distractions. Varying learning techniques provides additional novelty that can help raise attention. For example, create game-like conditions; change the context often; make the message as personal as possible; all increase attention.

Generation. Information is not stored in the brain as discrete memories, like a hard drive. Instead, memories are made up of a vast web of information from across the brain and are all linked together (Davachi & Dobbins, 2008). The more associations, or entry points linked to the original information that are connected to a memory, the thicker the web will be; therefore, the easier it is to find a memory later. Using metaphors and stories creates more associations.

Note: despite being widely thought central for learning, research has shown that repetition has only a limited impact on creating lasting learning (Wozniak & Gorzelanczyk, 1994). So, what works? *It depends.* Psychological and neuroscience research have shown that the key to optimize learning and building a long-term memory is to create personal ownership of learning content. This ownership or generalization of students' own learning occurs when an individual is motivated to understand, contextualize, retain, and apply knowledge in their own way, and in different environments that the information was first gained in. Therefore, students and other receivers of information should be encouraged to take in the presented information and personalize it by transforming it in ways that are meaningful for them. This act itself creates a rich set of associations activating the hippocampus.

Emotion. Learning happens in many complex layers with emotions being one of the most important regulators of learning and memory formation in the brain's hippocampus.

The way in which emotion is thought to enhance memory and learning is twofold. First, emotions grab the attention of the individual and therefore help to focus attention on the event that was the initial stimulus. Second, the hippocampus sits adjacent to the amygdala which helps to signal the hippocampus that a particular event is salient and thus increases the effectiveness of the encounter.

Of course, one challenge of using emotions in a learning context is the difficulty of creating emotion-arousing events that are positive. It's more common and easier to stimulate negative emotions such as fear and threat. It is well known that pleasurable anticipation and positive feedback leads to an increase in dopamine neurotransmitter that helps learning stick.

It can also be useful to have a training structure that includes novelty and entertainment; this normally stimulates positive emotions for the student.

Spacing. It has been shown for some time now that distributing learning over time is better than cramming learning into one long training session. Training in long sessions can increase short term performance, however distributing learning over time leads to better long-term memory, which is the ultimate aim of learning.

For example: Instead of training in one place for two hours, break training into several 20-minute sessions while you change the context at the same time. For example, when training alignment, do it in several different environments, the green, the tee, the bunker, the fairway, etc., for short time spans. This suggestion also holds when learning book content; change locations, do not stay in the same location for hours.

NOTE: Repeated testing is superior to repeated training for the formation of maximum long-term memory. Frequently put students through activities that test their skills is a suggestion that studies into the nature of learning make. Avoid having students just train, without frequent testing of the skills to be learned.

Summary

Learning and Performance are the consequences of memories being formed, reconstructed, and retrieved from brain network circuits. Each movement of a performance requires the coordination of sensory and motor systems in the peripheral and central nervous systems, in context of the prevailing environment: social, mental, and physical. Learning prepares you for a Performance. In general, then, there are DO OVERS during Learning (trial-and-error), but not during a Performance, you can't un-cook an egg! As will be discussed in Chapter Four, Learning and Performance are affected differently by Stress and Emotion. Factors that affect Learning are not interchangeable with those that affect Performance. As such, a Brain-Compatible Environment for Learning is not necessarily the same as for Performance.

Your Notes

Is All Learning the Same?

It Depends

"Bodily exercise, when compulsory, does no harm to the body.
But knowledge, when acquired under compulsion, obtains no hold on the mind."
—Plato

Hebron

LEARNING: Our brain is an information processor that has millions of wire-like connections that hold information. Acts of training and learning change these connections. I like to tell students with a smile on my face, I am coaching their brain and emotions not them.

From that view, all new learning is the same because there has been an internal change in a brain connection, or perhaps several connections. I suggest keeping in mind that each student is different, and any methods used that supported deep learning were emotionally safe and compatible with the brain's connection to the nature of learning. **But there are variables.**

The aim is to guide students in the direction of developing their own personal tools for learning any topic. I explain they have the capability to change, and I will not provide the exact answer but guide them in the direction of finding their own way. I call this "out-struction," with the solution coming out of the student.

When I am with students for the first time I say, "I cannot teach you, but I can help you learn." For example, roommates could help each other get through law and medical school, but you cannot teach someone to be a doctor or lawyer. We can be shown a car steering wheel, brake pedal, and gas pedal, but no one can teach you how much effort to apply when using them. These skills can only be self-discovered. I refer to this as POSITIVE FEAR

Hebron

Schmidt and Wrisberg (2000) suggested that when learning, transferable elements could be categorized into:

1. **Movement** skills—biomechanical and anatomical actions.

2. **Perceptual** skills—environmental information that individuals are interpreting emotionally.

3. **Conceptual** skills—strategies, guidelines, rules.

Sports-skill demands include:

1. **Physical** demands, such as power.

2. **Movement** demands, such as precision and aesthetics.

3. **Cognitive** demands, such as perception, memory, or strategic capabilities.

These are developed more efficiently through deliberate play, than structured deliberate practice.

"Tell me and I will forget, show me and I will remember, involve me and I will understand"
—Confucius

Yazulla—Conceptual and Motor Skills

Despite many similarities, Learning differs for conceptual and motor skills. These differences are critical for Learning in classroom and coaching environments. As mentioned, all learning depends on rewiring existing patterns of brain circuits to form new or different memories. During learning, some brain connections are a) weakened or removed, while others are b) strengthened, or c) new ones formed during learning. New connections hold new information. A major difference is the amount of sensory and motor systems that are involved in the activation and execution of these brain circuits.

As in the example of learning a poem, in general, the more you involve multiple senses and motor behavior in the learning situation, the easier it is to form long-term memory. When multiple neuronal circuits that hold dissimilar information (size, color, shape) are simultaneously activated together, they form stronger connections in the brain. This is also true for rehearsing or repetitions. Overall, academic classes require little physical activity. They are largely conceptual, requiring memorization, critical thinking, abstract thinking and creativity, for example: physics, philosophy, economics and music theory.

Hebron

Contrast these with sports, dance, and music performance. Here, conceptual learning is necessary with the added complication of coordination with the motor system. The training and integration of sensory-motor networks is the task of sports and theatrical coaches. Motor skills both open adaptive skills like soccer, and closed planned skills like typing, and cognitive learning, while different, have some internal similarities. Both are influenced by an individual's motivation, emotion, self-image, self-evaluation, among other relevant considerations including memory and past performance.

Motor-skill acquisition is the process by which movements are learned and produced alone, or in sequence, and come to be performed effortlessly over time through interaction with the environment. First, learning the task involves incremental acquisition of individual muscle units into a movement. Second, that movement becomes coordinated with other individual movements forming a pattern. Those patterns then adjust to different conditions. For example, the same muscle movement is used to lift a coffee cup, regardless of whether the cup is large, small, empty, or full. The difference is the effort involved, and your adjustments based on your expectations and experience. The adjustments in motor patterns are possible because connections among neurons in the brain are continually being modified; it is not an exact act.

The phrase "use it or lose it" certainly applies to the persistence of what is referred to as "muscle memory." **The muscles do not have a memory as such.** The memory is contained in the brain circuits that coordinate the muscle movements of motor behavior. Each brain cell and their connections contain information, which is exchanged with other neurons throughout the brain constantly.

Multiple brain areas are involved in the use of any movement, not only the particular movement, i.e., picking up a cup, but also whether to move, when to move, to start the move, how far to move, and so on.

Skills and mental pictures and thoughts are activated for desired outcomes. Hopefully, skill activities have been organized in ways that reduce mental demands so that performers can direct their thoughts to other features of the activity, such as strategy. While a motor performance can be observed, motor learning in contrast is an unseen internal process that reflects the performer's mental capabilities for producing a particular task externally.

Memories of past or expected events, involving fear and doubt, play a significant negative role on the internal ability to learn and the external ability to perform. Learning and performing are brain (mind)—body events. **The brain *knows,* but it is the muscles of the body that *perform.***

A traditional coaching technique is the use of learning aids either hands-on or some mechanical device. When learning a golf or tennis swing, the instructor or an aide may try to physically guide the student through the desired motions. These aids constrain or externally control the movement of the club, legs, arms, hands, etc., and remove struggling, which can be the most useful tools for learning. It's been found that self-defined movements are more resistant to forgetting than when movements are physically controlled by others and aids.

Yazulla

One reason for this is that our brain's control of movement depends on feedback from the muscles and joints that are moving. The information, supplied by receptors in your skin, joints and muscles to the brain, is called 'proprioception' and 'kinesthesia.' Proprioception provides information to your brain where your body parts are in space relative to each other, even with your eyes closed. Ever wake up in the morning with your arm or leg asleep, you do not know where it is, and you cannot move your fingers or toes. That is because the reduction of blood flow to your arm has cut off sensory information from your arm to your brain. Not only do you lose sensation, but the limb is paralyzed. Once blood flow increases, after the pins and needles pass, position sense and movement return.

Kinesthesia provides information as to how your body parts move in space. Examples are touching your nose, bringing your fingers together, lifting a cup of coffee to your mouth, and so on. In order to touch your nose, you must know where your hand is in the first place and then direct it up or down, right or left, etc. to your nose. If someone else just passively moves your arms to take you through a golf swing, you are not actively contracting your muscles. Sensory information from your joints will tell you where your body parts are (i.e., vertical *vs.* horizontal, etc.).

However, the passive movement removes much of the kinesthetic feedback from the muscles contracting and associated stress on the joints that your brain uses to assess arm movement, position, and subsequent control. It is the multisensory input and feedback from the skin, joints, and muscles following active motion that are required for coordinated motor movement as coded in the brain. *To repeat—movement is encoded in the brain NOT in the muscles.*

Do learning aids have a place? Yes, but with constraints. Alignment sticks are used even by pro-golfers during tournament practice to calibrate the body position in relation to the target. A weighted club head or baseball bat is used in hopes of increasing swing speed. However, constraining body position with braces during a movement (like a golf swing) may not be very helpful because the compensatory movement of opposing muscles is critical for proper coordination of any movement. This is why lifting free-weights is more effective than using a weight machine. The weight machine reduces the need for the opposing muscle activity needed to maintain balance.

Hebron

It is faulty to believe that you will one day have perfect physical technique. Studies at the Electrical Engineering Department of Stanford University on the neural basis of sensory motor integration and movement control has revealed that our brain simply does not allow us to 'code' the swing perfectly, so as to repeat it time after time. We all have inconsistent swings. The main reason you can't move the same way each and every time, is that your brain can't plan the swing motion the same way each time. It is as if each time the brain plans how to move, it does it new and differently each time. Practice and training can help the brain guide our motion, but people and other primates simply aren't wired for consistency like computers or machines. Instead, people seem to be trial-and-error improvisers by default (Shenoy, et al., 2006).

It was thought that the inconsistency in a repetitive motion, for example: golf swing, shooting baskets, throwing a ball or darts, resided in the muscles. This is only partially true, accounting for about 50% of the variability. The output of individual motor units in any movement may change with each repetition, perhaps due to fatigue. The rest of the variability is in the brain, the command center for starting the movement. An enormous amount of neural coordination is required among numerous brain regions for even the simplest motion, such as wiggling your finger.

Triggers Into Action 1–2–3

1) The intent to move prefrontal/premotor cortices, 2) This triggers a sequence of events involving the motor cortex, subcortical areas, and cerebellum, among others as a final signal is sent to the spinal cord, 3) whose axons activate the muscles. A change in any of these steps will affect the input to the muscles, and consequently the resulting motion. **There is novelty in every movement.** The best we can hope for is to be in-the-ballpark, with a scatter as close around the target as we can manage.

When it comes to learning, there are serious questions about what Computers and TV replay actually provide that can support learning. But because of tradition, there is some resistance to reflect on how useful or not this kind training is to learning. Developers of feedback systems have said they are using that technology because it is available and not because they have researched into what is best for the user, for me a curious observation.

These developers of motion-analytical programs appear to have little cognitive research about how much, or what type of feedback is appropriate, or when this kind of feedback should be delivered; or even if it may be detrimental to progress—An interesting thought.

Yazulla

What about feedback? The concept of feedback is that the result of every action is used to adjust the sequence of a subsequent action. One example is a thermostat. Once a room temperature is set, any deviation from that set point triggers a response in a heater or air conditioner to adjust the temperature to the set point, at which the system stops until another deviation is detected. This is negative feedback that is used to keep a system operating within some limits. Another example is walking with a full cup of coffee, or mug of beer, that requires continual sensory motor feedback to maintain balance and the integrity of the cup/mug contents. However, if you are jostled, knocked off balance, negative feedback would have difficulty making the appropriate neuromuscular adjustments to restore balance. Even if you could maintain your balance, you could not stop the movement of the coffee/beer in an uncapped cup, and the liquid would most likely spill.

Regarding a golf swing or any very rapid movement, negative feedback in the nervous system is too slow to make adjustments once the movement is on its way. For a golf drive, the ball is stationary, perched on a tee; everything depends on the ball staying still. Once you start the downswing you normally would not be able to adjust the swing if the ball moves. Your reaction time simply is not fast enough. This is different for a putt, in which the golf club movement is much slower than it is for a drive.

Similarly in baseball, a batter tries to predict the final location of the pitched ball when it reaches them. A checked swing can be made if the ball is not too close yet. However, if it were possible to adjust the swing as the ball got closer (within 15 feet), no batter would strike out. A 95-mph fastball is traveling about 140 feet/sec, covering those 15 feet in about 107 msec (about one-ninth of a second). This is far too short a time for a human to visualize the ball, calculate its trajectory and make the sensorimotor adjustments in order to correct the swing.

An adjustment for movements faster than the controlling mechanism is relegated to a Feedforward plan for a future event, not the current event. This is how the brain works to control and plan rapid movement. All elements of the movement (i.e., golf swing) are analyzed: body parts involved, ball position, terrain, obstacles, distance, direction, club selection, etc. The full brain is involved, visual, hearing, balance, skin senses, premotor cortex, motor cortex, subcortical areas, cerebellum and so on. The information is integrated into a motor pattern of muscle group activation.

The fine tuning takes place in the cerebellum. The final output of the brain is transmitted through the spinal cord to motor neurons that then activate the skeletal muscle. The motor neurons, themselves, receive tuning type inputs from the peripheral tissues muscle, joints, tendons, and other neurons within the spinal cord. All of this information from brain, spinal cord and peripheral tissues is integrated at the motor neuron, then, signals are sent to activate the muscles that form a particular muscle group.

Now you take your swing. The result is analyzed and compared with the intended outcome. The new information is used to prepare another program of motor movement to duplicate or adjust the movement to a desired outcome. The point is that feedforward depends on trial-and-error outcomes that look to the future, not the present (Dowrick, 1999). Each subsequent movement is activated by a motor circuit that has been modified by analysis of the outcome of a *previous* event to plan a *future* event. On top of the motor circuits, the emotional response to each outcome brings in other areas of the brain to affect an already complex motor system. In such a complex system, robotic repetition is extremely unlikely. Emotion and stress were treated in Chapter Two.

Hebron

Side-by-side comparisons of a student to an expert can be damaging emotionally and to memory formation. Immediate feedback is not desirable or necessary; and it can act against the development of performance. Given the strong agreement among motor-learning experts and motion-analyst scientists that there is no such thing as optimal movement patterns, trying to copy a performance **exactly** has been shown to reduce personal skill. Extrinsic feedback from an outside source often causes confusion and does not necessarily improve performance.

Suggestion: encourage the use of **guided** self-discovery learning environments that support individuals through different possible solutions. This minimizes the possibility of information overload. Instruction has often given students OVER-PLANNED LEARNING EXPERIENCES that are overlooking that less-technical details normally are more supportive of the nature of learning.

I now see myself more of a Learnest, guiding student learning and development, than as a teacher who just provides information. How information is shared with or obtained by students is critical to their growth and development. That words can support or suppress learning is a reality that often gets overlooked.

Information from several fields of science caused me to question the value of some long held and widespread assumptions that supported get-it-right methods and approaches to teaching that I was using. For years I was overlooking that confusion and struggle are welcome additions to the nature of learning in emotionally safe, learning environments. Perhaps a counterintuitive view, but true. Having students make mistakes during learning and training is a brain-based learning approach. Poor outcomes become reference points for what to do differently. A very USEFUL insight.

> *"Every person has two educations, one which he receives from others,*
> *and one, more important, which he gives to himself."*
> —Edward Gibbon

Summary

All learning and performance require: 1) input of sensory information, 2) integration in the brain, and 3) coordinated output to the motor system. Differences are due to the relative amount of input information from any of the senses (sight, hearing, touch, etc.) and the degree of motor output involved in demonstrating that learning (speech, painting, golfing, ballet, etc.) has occurred. The fundamental mechanisms underlying all learning are the same. These mechanisms are also subject to the same influences of fatigue, stress, talent, motivation, emotions, and past experiences.

Your Notes

CHAPTER FIVE

How Does Motivation Affect Learning and Performance?

It Depends

"We are generally more effectively persuaded by reasons we have ourselves discovered than by those which have occurred to others"
—Blaise Pascal

Hebron

MOTIVATION: When I first meet with a student it helps me to gain insights into why they play golf and want instruction. People of all ages and backgrounds come to golf for different reasons and goals. Remember, I have coached individuals who just play for outdoor exercise; others who play for social reasons; some who play because they are competitive with themselves; some who play or want to professionally; some because their parents want them to play; others who are hoping to gain a scholarship to college; some who play to spend time with family members. I could go on, but people are motivated to play golf for different reasons and their ability to learn and develop their physical and mental skill is heavily influenced by why they are playing. While physical skills are required to perform, it also depends on the mindset and emotions of the performer.

Internal and External Motivation

Yazulla

Motivation to learn and perform can derive from internal and external sources. Positive internal motivation appears to develop from an intense internal need to learn and achieve, rather from negative external pressures for success. Examples include simple curiosity, artists, professional and

Olympic athletes, CEOs of large corporations, leading academicians in any field, experts in any trade, etc. External sources of negative motivation include peer pressure, parental and family pressures, coercion, etc. Except for curiosity, it is possible that some internal motivations develop from youthful experiences of being exposed to a variety of activities, one of which may have triggered an intense interest, for example, sports, music, history, science. Regardless of the source, motivation is a powerful factor, negative or positive, in learning and performing.

There is a long history of research showing that motivation can be altered by adjusting a Reward or Punishment. The fear of failure is a particularly powerful motivator, for action as well as inaction. There is always the conflict of when to act, when not to act; the conflict of curiosity and self-preservation, as illustrated in the following pairs of apparently contradictory aphorisms:

He who hesitates is lost vs. Look before you leap

Strike while the iron is hot vs. It is easier to get into a situation than to get out

People are more willing to learn and perform for a large payoff than for a small payoff. Similarly, they will work harder to avoid a large punishment than a small punishment. The conflicting motivations are weighed and eventually a choice is made, with all the consequences, some foreseen and others unforeseen, as in the *law of unintended consequences.*

One's behavior in a defined activity often progresses in a sequence of steps from immediate, short-term to longer and long-term goals. For example, a goal for today may be to get a good grade in an exam, with a goal to getting a good grade in the course, with the goal of a high grade-point average, with the goal of getting into medical school, graduate school or a job offer, with the goal of succeeding in whatever choice was made, and so on. There is a similar sequence of goals for learning golf: stance, swing, accuracy, distance, score, etc. The motivation for each goal attempted, achieved, or not, only leads to attention on another goal. The emotional reaction to the outcome of any effort either advances or retards one to achieving the goal. The emotional reaction to the outcome also depends on the perceived reward or punishment associated with the outcome.

The result of not achieving a goal is either to re-evaluate the merit of the goal (*is it worth it?*) or try again *drop back 10 and punt.* Whether or not the goal was met depends on the demonstration of learning in whatever performance was used to evaluate the learning. Here the quality of learning is not necessarily correlated with the performance. **Successful learning does not guarantee successful performance.**

All learning is based on memories formed by reorganization of brain circuits in a complex mixture of chemicals and hormones. Anticipation of a large reward or punishment causes the release of Stress Hormones discussed in (Chapter Two) that increases blood pressure, breathing and heart rate in addition to their effects on the brain. The emotional load of a negative motivation "Learn this or else" depends a lot on what "or else" is. Conversely, the positive motivation of "Learn this and you can do or have this" depends on the magnitude of the reward. Regardless of the positive or negative motivating factor, the distraction caused by high anticipation, along with global hormonal effects, will make it difficult to concentrate and learn the task at hand.

Motivation and Learning

Yazulla

A college academic setting has its own peculiarities that differ from earlier grades. Often, the first question asked of a college freshman at orientation is "Why are you here?" The answers fell into three general categories: a) biding time, b) another's expectation, and c) goal directed. As expected, the motivational level for each of these groups is quite different so *it depends.*

In general terms, those biding time perhaps are just putting off getting into the job market, not sure what career path to take, just see it as a natural extension of high school, look at it as a time away from home, party time, and so on. The motivation of many biding time students is simply to not flunk out. If I flunk out, it is then back home and "then, what now chief?" I have a decision to make. The goal then is to do just what is required to get a "C," or in my day the "gentleman's hook." With this goal, the motivation to acquire an education, that is, obtain a critical understanding of the material, takes a back seat to the just passing grade.

Students whose motivations are to fulfill someone else's expectation or are internally driven are in a similar and difficult position with respect to stress. For many of these students they were the first in their family to go to college, and the expectations from the family are high. The goal is the *Grade,* and that grade is an "A;" nothing less than an "A" will do. Admission to medical school is very competitive and a "B" in a science course can severely reduce one's hope of getting into a U.S. medical school. The pressure on these students was intense. Of course, there were many talented students who worked hard, did what they had to do and more, and did extremely well. For those who found things more challenging, extreme motivation often produced desperation and the strategies developed for success at any cost ranged from excessive cramming to cheating.

The best example of excessive cramming is the All-Nighter, a right-of-passage for many freshmen. One simply cannot cram a semester's worth of information into one or even two all-night binges. I know. I did it once and got an "F" in a weekly physics exam. Lesson learned. As mentioned in Chapter One, deep Slow Wave Sleep is important for the rehearsal and retention of memories formed during the day. Without this deep sleep, there is less chance for retention of material that is read during a binged all-night study session.

The effects of sleep deprivation on brain chemistry, hormonal function, alertness, memory recall, attention, etc. are well known and experienced by anyone who has stayed up all night and tried to function the next day. The shift from daylight savings time to standard time and back again, even only by 1 hour, affects many people for the worse. The health of workers on night shifts and even worse, on rotating shifts, is known to be adversely affected. The best preparation then for an exam is a good-night's sleep.

Despite the differences in the individual goals and strategies to achieve high grades, amazingly, the outcome was that the grade distribution in the class did range from "F" to "A," and all in between. Ability aside, the reason boiled down largely to differences in motivation and effort the students were willing to expend in the course. I often found that students would perform simply up to their own level of expectation and no more. I expect that this attitude of "just good enough" would carry over to other aspects of their life. Perhaps an activity that really caught their interest would inspire them to greater effort to actually do their best. For many of these students, the payoff to earn the Grade simply was not worth the effort to achieve it, and they were content to just settle.

The other strategy for those who were less motivated or had a great fear of not making the grade was cheating. Early on in the 1970s and 1980s, cheating was largely bringing notes into the exam or copying from a neighbor. With time, cheating became more sophisticated, with the use of cell phones and ear buds. Competition to the point of sabotage of laboratory reports and cheating on exams just illustrated the consequence of the motivation to Perform without regard to Learning. With the considerable effort devoted to methods of cheating, Learning and Performing are dissociated. Unfortunately, the smug attitude of "I got away with that" replaced the satisfaction of a "job well done." The quote below by Andrew Carnegie seems a bit harsh, but is unfortunately, true, as exemplified as in the underachiever.

People who are unable to motivate themselves must be content with mediocrity, no matter how impressive their other talents."
—Andrew Carnegie.

Hebron

In my view, at every level of golf skill, professional and amateur, many students do not recognize that learning environments, instruction environments and performing environments are not the same. Many club golfers lack patience and do not recognize the time it takes to develop different skill sets. Swinging a golf club at the range is much different than when playing from different lies on the course. Factor in sand bunkers, chipping, pitching, and putting, the game of golf requires on-the-fly adjustments on the part of the student. Unlike in academia, in which the teacher can give the answer, in golf, the coach cannot swing the club for you, can only give advice.

Outcomes in my mind are the student's real educator and I am their coach or guide. The reality is I do not actually know what the unintended or workable outcome felt like to the student. Also, I do not know what the student actually did mentally to produce the outcome. Yes, I did see the external body and club motion, but cannot see what was going on internally to create the outcome. **You cannot film the cause of the outcome, only the result.**

To motivate student learning I ask questions that promote and guide student introspection. After the outcome I may ask some questions that include:

- What did you do?

- What did you think of that outcome?

- What was different for you that time?

- What was your picture?

- Was that faster or slower than last time?

- If you were the instructor, what would you say about that outcome?

- What would you suggest for making that outcome different?

Motivation and Performance

Yazulla

Compared to Learning, the role of Motivation on Performance is more nuanced. Consider two extremes: performance can be 1) low physical output of mental learning, or 2) high physical output of motor learning. Consider low physical output of mental learning, regardless of how well something was memorized, a musical piece, part in a play, historical events, chemical formulas, math concepts, legal cases, etc., extreme motivation, whether positive or negative, will likely affect performance for the worse, at least at the beginning of the performance—"Butterflies." Often, once the initial stage fright is over and they get over the hump, the person will build up a head of steam and proceed successfully. In sports, that person is said to be "good in the clutch." Conversely, if a roadblock is hit and the person does not get by; it is all over. At such times, that person is said to have choked.

In contrast, high physical output of motor learning would refer not only to sports but also to any kind of labor involving tools, manual or power. High motivation, by releasing stress hormones, prepares the body and mind for heavy physical activity. These activities use the large muscle groups for running, jumping, and throwing. For example, in sprinting, any race, shot put, javelin, weightlifting, etc., full effort continues to the end of the activity.

However, in many activities, at the last minute there is the requirement to shift from large motor to fine motor control, for example, biathlon cross country skiing to rifle shooting, a game winning field goal, a foul shot, 3–2 pitch in the 9th inning, hockey or soccer penalty shot. All activity stops, there is no more running or jumping, just one person standing alone with the outcome of the game on their shoulders. The Motivation to succeed is increased by the burden; the stress hormones may be in full force. However, what is required is a calm directed activity, despite a pounding heart. The motivation?—a chance to be the hero or the fear of failure, letting everyone down—probably both. It is the ability to perform under stress that can largely determine the outcome, for better or worse.

Hebron

An often-asked question of golf instructors is, "How long will it take to improve? Of course, *it depends*. People play golf for different reasons: Common goals include for personal pleasure and satisfaction, socializing by joining a league, or entertaining clients for business. Each of these goals has its own levels of motivation and stress associated with it.

The motivation to learn can be similar, but the reason for learning is often different. All students want to learn how to golf but making the most out of time available is the bridge to cross. Are students hearing positive messages that support motivation, confidence, and development, or are the messages judgmental, critical and do not motivate or create self-confidence? *It depends*.

In my view, confidence and motivation do not come from the student but develop from what they hear and receive from others. Also, people who take up the game with others, or have an ongoing date to play with others seem to learn at a faster pace than those who do not. Perhaps, it is the motivation of peer-approval and obtaining such that has a positive effect on learning and more importantly, performance. One may get a great deal of personal satisfaction from scoring an 'eagle' or a 'hole-in one,' but it rings a bit hollow if you were alone and there was no one there to witness it or share it with back at the 19th hole.

> *To get the full value of joy you must have someone to enjoy it with.*
> —Mark Twain

What About Praise?

"Neither worse nor better than is anything made by being praised."
—Marcus Aurelius

Hebron

Praise is distractive/Praise is productive. Both statements are true, so *it depends*. But the first one is based on evaluation and the second one is based on appreciation. Hopefully, the words said to students after an outcome are used to recognize what worked and was appreciated about their efforts and accomplishments. The student then draws their own conclusions about the value of the outcome and themselves. After appreciative words, the students' conclusions about themselves can be positive and productive.

Avoid judgmental praises such as "you did a good job" or "you are a hard worker." Using these statements can cause students to develop their self-image based only on what others say and not on their own feelings. In psychotherapy, individuals are not told "You are doing great," or "Keep up on your good work." WHY? Judgmental praises are avoided because they are not helpful, can create anxiety, invite dependency, and evoke defensiveness. A judgmental-praise from others is not conducive to developing self-reliance, self-direction, and self-control. Development of these qualities demands freedom from outside judgments. They require reliance on the student's inner motivation.

When students are self-reliant, they are free from the pressure of judgmental praise. Praise can make students feel good for that moment, however it creates dependency. Others become the source of approval. Students own self-worth then depends on others. "You just did that; how do you feel about that?" This statement invites students to make their own inferences. "Is that the outcome you wanted? Praise without using valuation adjectives is the aim. Trying to please others can suppress learning, development, and growth.

Yazulla

I agree, praise is a double-edged sword. Dependence on the praise of others, as an external motivation, can lead one to define their self-worth on pleasing someone else. The reality is that humans are social creatures and can take great pleasure in accolades from family, friends, colleagues, and co-workers for a job well done. This is obvious as indicated by the popularity of award shows, game shows, sports championships, recognition in the sciences, arts, community, and so on. Human society thrives on praise and is terrified of negative criticism.

Praise is also critical for the socialization of children. Behavior according to social norms is better achieved by positive reinforcement (praise) than negative reinforcement (punishment). Smiles and hugs are very effective in promoting desired behavior all the while instilling a sense of wellbeing in the developing child. Eventually, one hopes that motivation for the child shifts from pleasing others to personal satisfaction for their proper social behavior. As mentioned earlier, negative reinforcement may be required to deter children from dangerous situations until they are mature enough to understand the danger. Society has ways to show its displeasure for asocial or antisocial behavior; exclusion is a very powerful tool to enforce social behavior.

Praise for any activity that one does, regardless of the outcome, often rings hollow. When children are very young, medals awarded just for participation may be cute. But eventually such medals become meaningless, or at least they should be seen as such as the children get older. When all actions are praised equally, over time there is no incentive to improve. When approval and praise are routinely offered, pleasure and the emotional high decreases. Most everyone enjoys an "attaboy," particularly when deserved. But deep down, people realize the value of that praise in context of what they actually accomplished.

My favorite example comes from my granddaughter after an elementary school band concert. My wife told her that the performance was wonderful, just great. My granddaughter replied *"Grandma, we stink, you just said that because I was in it."* Here was an honest appraisal of a perceived outcome by a youthful participant. No praise, however well intentioned, was going to alter her perception.

And so, we are back to the double-edged sword of praise. Humans are social creatures and would not function or thrive in an environment of robotic apathy in response to their actions. Approval and praise are critical for the social development of youth as well as for the smooth interactions among individuals. However, when the motivation for any action is praise, that person surrenders their self-esteem and sense of self-worth to the opinion of others. It is far better to be comfortable in your own skin, do your best and accept any praise gracefully.

> *"Dignity does not consist in possessing honors, but in deserving them"*
> —Aristotle

Hebron

Wrzesniewski et al. (2014) studied motivation in West Point Cadets, particularly the role of two types of motivation, internal and instrumental. For example, conducting research for the joy of uncovering facts is an act motivated by internal motivation. Conducting research to become well known or wealthy is an act motivated by instrumental motivation. Some believe mixing these motives would be a good thing, but it is actually counterproductive.

The authors suggest efforts should be made to design and structure learning activities so that instrumental consequences do not become motives. Helping people see the meaning and the impact of their efforts, rather than gaining external rewards, may be the best way to improve not only the quality of their work, but also their success.

This paper shows when students are not that interested in learning, external incentives may prompt participation, but can result in fewer well-educated students. Trying to make an activity more attractive by emphasizing both internal and instrumental motives is understandable but can have the unintended effect of weakening what is essential to success, internal motives.

Over 11,000 cadets in nine entering classes at the United States Military Academy at West Point were questioned about their motives for entering the academy. Cadets with strong internal and strong instrumental motives performed worse on every measure than cadets with strong internal motives but weak instrumental ones. They were less likely to graduate, less effective as military officers and less committed to staying in the military.

It may be useful to note that a person can perform a task well that may have both internal and instrumental consequences. For example, trying to learn (internal consequences) to get good grades (instrumental). But just because an activity can have internal and instrumental consequences does not mean that the people who thrive in these activities had motives that were both internal and instrumental.

There is a temptation among educators and instructors to use what motivational tools are available to improve performance. While this strategy may create participation, it detracts from actual student learning.

Summary

Motivation, regardless of whether it is internal or external, negative, or positive, has an effect on the functional state of the brain and body. Learning and Performance do not occur in a vacuum, but in the internal environment of the brain and body along with the social environment at that particular time. A motivated person is a stressed person. As discussed in Chapter Two, all stress is not the same: stress may be helpful or harmful. There is no doubt that a positive motivation of some reward is more effective for learning and performance than a negative motivation of punishment or ridicule. A motivation of praise is life-long but should be kept in perspective for what it does to one's sense of self. Praise, honestly earned and sincerely offered, is much appreciated—but don't depend on it.

> *"Care about what other people think and you will always be their prisoner."*
> —Lao Tzu

Your Notes

CHAPTER SIX

Are Unintended Outcomes Failures?

It depends!

"A life spent making mistakes is not only more honorable,
but more useful than a life doing nothing."
—George Bernard Shaw

Hebron

FAIL = **F**inding **A**ccess **I**nto **L**earning—A positive View

FAILURE—In the past, during instruction and playing golf, my students saw poor unintended outcomes as failure. A negative view that I also had at that time. But today I point out to students who are not happy with their ball flight outcome that their swing motion was perfect.

That ball flight was the only possible outcome for the impact conditions that their motion of the swing created. But it was also an outcome to learn what to do differently.

Today I point out that errors are useful. Errors and struggle are our most useful learning tools.

"A man's errors are his portals of discovery."
—James Joyce

Yazulla

I think the notion that "there is no failure, only another opportunity to learn" needs to be qualified. There is a big difference between an unintended outcome (failure) during a learning situation and during a performance situation. According to the Merriam-Webster 3rd International Dictionary, there several meanings of the noun "failure," none of which are positive:

1. omission of occurrence or performance; failed to show up for a meeting

2. an abrupt cessation of normal functioning; washing machine failed

3. lack of success; "messed up"

For our purpose, number three, lack of success is a more relevant definition in that it implies a value-loaded term that speaks to the perceived outcome of any action. The word "failure" has a negative connotation, whereas unintended outcome is a softer term. There are do-overs during learning but not during a performance. However, both can serve as learning experiences.

Before outcomes are discussed, it would be useful to discuss choices/decisions and the conditions under which they occur. An outcome is the consequence of a choice that is acted upon. A decision is required when an organism is presented with a situation that can be responded to with alternative actions: hunger, thirst, defense, etc. When a choice is made, not only must we consciously activate the neural circuitry for the chosen action, but the neural circuitry that would activate the unwanted or opposing action is unconsciously suppressed. A simple analogy is the braking of a car. When you apply the brakes, you take your foot off the accelerator.

Conversely, when you want the car to go, you release the brakes and press the accelerator. The result of doing both at the same time depends on which action is the stronger. It is extremely inefficient to activate two competing processes in order to get anything done. The choice to do something, then, always involves, unconsciously, the choice not to do the opposing action. You cannot move something if you pull and push it at the same time.

Virtually every situation we encounter is novel. The physical context may not change but the cues do. Context is the unchanging part of the environment: tree, house, etc. Cues are variable: smells, sounds, breeze, people present, conversation, day, night, temperature, etc. We live in an orderly universe in which events follow from cause and effect. We depend on this consistency in order to make predictions, decide and then act. Before a decision is made, we play out a prediction scenario, for example, cost/gain, present/future, odds of success/failure; if this/then that. Once the decision is made and acted upon, the outcome is observed and analyzed. Regardless of whether the outcome works or is worse than expected, we revise the prediction to modify our action to improve the probability of success with a different outcome.

To do this, you must remember what you did in the first place and then reproduce or alter it given the constancy of context and changing cues. Any revision or adjustment of your actions is a learning process. Both elation of success and disappointment with failure are associated with emotional and physical responses that are similar in some respects: Increase heart rate, flushing of skin, physical displays joy *vs.* disgust, fist pumping *vs.* fist punching. When unintended outcomes are accepted for what they are and not judged or criticized, learning is emotionally safe, with a student-always-first environment mindset.

The effects of the stress hormones on Learning and Performance in these situations were treated in Chapter Four. It often seems that our learning situations consist of an endless succession of outcomes and revisions. In this regard academia and sports differ. No one is expected to reinvent the Calculus or discern the Double Helix of DNA from scratch. In the sciences, students replicate the past as a learning experience, but do not rediscover it. In sports, one learns much by play, even the fundamentals. Give a child a soccer ball and tell them to play with it for a month or so no hands and then have them show you what they can do with it. I expect they will display respectable competence.

However, give a child a sphere and ask them to come up with the formula to determine its volume mathematically [Vol = $4/3\pi$ radius^3], and it is likely that they will not be able to do that. The point is that there are some activities for which play-to-learn is appropriate and others, for example solid geometry and calculus, for which you must stand on the shoulders of others who help you learn. It is a joy to experience an "ah–ha" moment of insight, whether on your own or with help.

> *Our failures are more instructive than our successes.*
> —Henry Ford

Our culture is imbued with the notion that winning isn't everything—winning is the only thing. This attitude places enormous value on winning and contempt on losing. There seems to be a built-in neurological response to unintended outcomes, disappointment that can lead to frustration. The goal of education, therefore, seems to be changing the attitude towards a mistake or an outcome that is less than desired. Here is where the error as opportunity comes in. Rather than being berated for falling short, one should be encouraged to try again. I suppose this is the attitude in horseback riding—when (not *if*) you get thrown, get up and get back on the horse. Sometimes the silver medalist has a real grump on their face. Other times it is heartening to see a silver or bronze medalist glowing because they performed at their best. Knowing you did your best even though someone else did better is your reward.

Sad images from the 2020 Tokyo Olympics were Japanese athletes apologizing for not winning a gold medal. Kinichiro Fumita, the silver medalist in Greco-Roman wrestling, "I ended up with this shameful result. I am truly sorry." Sometimes your best just is not good enough; it is not your day. The ability to let it go and bounce back reduces the emotional baggage a person takes to their next effort. Dwelling on the past failure only adds to the stress inherent in any future performance.

"What seems to us as bitter trials are often blessings in disguise"
—Oscar Wilde

The point for a brain-compatible learning approach is that the neurological events that accompany unintended outcomes are built in, real and dysphoric. It is likely that there also is a learned component that is due to cultural expectations. The neurological changes in poor outcomes differ from those elicited by 'success' and likely have some adaptive advantage for survival in the short term. However, persistent poor outcomes, when called 'failure,' lead to emotional frustration that is neurologically and physiologically counter-productive. The goal then is to mute the emotional response to 'failure' by redefining 'failure' as the opportunity to improve and learn to do it differently, guided by the poor outcome. Realistically, you do not expect people to jump for joy after a flubbed golf shot or doing poorly on an exam. However, identification of self-worth with the successful outcome of a decision leads many to simply give up rather than try again.

In a learning situation, one often proceeds by trial-and error, which by definition is, adjusting your next step according to the outcome of the last step. Immediate success in a new task is often derided as beginners' luck, a tacit acknowledgment that it is only by repeated tries following unintended outcomes does one really learn. Unless we are dealing with one-trial learning as in a hot-stove example, all learning situations involve a series of steps that include intended and unintended outcomes!

It helps to see there is no such thing as a failure in an it-depends learning situation. Indeed, failure, unintended outcomes are necessary and opportunities in a learning curve. However, traditionally things change when we talk about Performance; that is where the rubber meets the road. We are not referring to a performance as in a rehearsal, sports scrimmage, or a practice examination. These are still learning opportunities. We are referring to a Performance when all is on the line.

In academia, that Performance usually takes the form of a test or exam, which is the barrier or hurdle to the next stage. There are many levels of "unintended outcomes" as one responds to their final grade in a course. A final grade of "B" or "C" instead of an "A" is a disaster if the course is Organic Chemistry and your goal is Medical School. That grade of "B" is not an opportunity to relearn the material for that student in that course. Rather, it may be an opportunity to reassess a career goal. For all other students, that grade of "B" may be a success or failure depending on the goal following graduation. In any case, it is called a Final Grade for a reason; there is rarely any recourse for the student except to evaluate what they do next. All of their effort to learn course content was boiled down to their ability to perform on exams.

"The gem cannot be polished without friction, nor man perfected without trials"
—Confucius

Hebron

My view starts with unintended outcomes that are valuable for learning what to do differently during both practice and performing. In golf, the distinction between Learning and Performance is clear-cut. During training or practice, one may hit 5 or 10 balls from the same spot, adjusting depending on the outcome of each shot. Trial-and-error is adjusting to the unintended outcomes, not seen as failure during training.

However, during a golf match, each golf shot is a Performance; there is no do-over, no Mulligan. The unintended shot may be a learning opportunity for a similar shot that may or may not occur for the rest of the round. But how one reacts to the unintended outcome is more important. Is it accepted as part of playing a game or does it create frustration that may last for the rest of the game? Remember—*It depends*!

Key components of the problem-solving process are question formation and exploration of options by students. Teacher-centered instruction is normally based on providing detailed information. In contrast, student-structured instruction is centered on a student's own thinking and problem-solving skills. Students are NOT GIVEN all the information needed in a nice, neat package–but must gather and sort out some themselves. **Guided** self-learning is an efficient kind of learning while training what to do, not what to fix.

Some instructors often go to great lengths to clearly define the problem a student will have to solve, how they should solve it, and point to the desired outcome. Students then do not have opportunities to engage with uncertainty, where learning starts.

Just as you cannot learn to swim if you never get in the water, students will not learn to respond productively to ever-changing conditions if they are never given opportunities to struggle. Instead of trying to eliminate the inevitable uncertainty with given details during instruction, promote problem solving and support ways students will come up with their own way of solving a problem.

It would appear that some golfers who are performing below plan are focusing on a fixing technique rather than on what to do. A key misconception is the BELIEF that you will eventually perfect your technique. Our brain simply does not allow us to 'code' the swing perfectly, so that it will repeat time after time. We are all in a sense doomed to a level of inconsistent swings. Inconsistent does not mean random. There is an expectation that the inconsistent swings and their outcomes will occur within narrowing boundaries as proficiency improves.

As we have stated: Unintended outcomes are a biochemical necessity during learning and have varying value during a performance. Learning a gymnastics routine, a piece of music or dance sequence requires errors and adjustments until a workable sequence is learned. An unintended outcome while playing golf offers other opportunities for growth. For example, learn to continue the performance with grace despite the unintended action.

What about 'intentional' unintended outcomes as a learning device? In the past, introducing some struggle during instruction was not recognized for its value. We now know that students learn more when instruction lets them wrestle with a problem. Struggling with a new idea, a different insight, a change in perception, creates changes in the brain's neural connections.

Jack Nicklaus, during his practices said he would deliberately miss-hit golf shots. This exercise would demonstrate the effects of misalignment of his stance, swing, follow through, etc. A similar outcome during a match would allow him to determine its cause and then try to do it differently.

The brain operates by making predictions, if this, then that. By having students repeat the unintended outcome several times, then asking them to do it differently to produce a different outcome helps students learn the more effective motion. They are learning important reference points.

For athletes, the inability to replicate the perfect movement might seem to be a frustrating problem that needs to be solved. But the brain has evolved an improvisational style of creativity precisely because the vast majority of situations require novel movements. For example, predators never get the chance to catch and kill prey in exactly the same fashion or in exactly the same conditions. Our ability to adapt is a survival skill.

For some there is the belief that we can 'groove' our swing to the point it just repeats and repeats while science is telling us the brain does NOT allow this to happen. If we accept that the swing will ALWAYS be somewhat variable from day to day, then training can take on a more flexible approach. Golf is a random game played in an ever-changing environment. Yet many players spend HOURS on the range hitting ball after ball in a fixed and closed environment. Perhaps more training time should simulate real game ever-changing conditions. People are good not because they are consistent; they are good because they adapt and handle inconsistency better than others. No one can predict their score on the first tee, and every swing produces a surprise.

"Experience is the name we give to our mistakes"
—Oscar Wilde

Yazulla

Deliberately introducing errors in a laboratory setting may or may not be a good idea depending on the consequence of the error. A deliberate error mixing food dyes to determine color mixing is no big deal. A deliberate error mixing ammonia and bleach just to see if what people say is true, despite the warnings on the containers, can be fatal. You can ask the class what would happen if you turned on an unbalanced centrifuge a device that spins test tubes at very high speeds, (in an enclosed area); but you do not want to demonstrate that.

Consider what an unbalanced clothes washer or dryer sounds like spinning at a few hundred revolutions per minute (rpms). Now imagine an unbalanced laboratory centrifuge spinning at 10,000–20,000 rpms. Although sealed with a door as the clothes dryer, the centrifuge would vibrate catastrophically and be destroyed, with parts flying all over. Therefore, deliberately introducing unintended consequences in a laboratory setting with chemicals, electronics, radioactivity, or power tools is most safely done hypothetically with schematics, questions, and scenarios.

Museum of Failure

Yes, a Museum of Failure actually exists! The exhibition has been on tour since 2017, appearing throughout the world. The curator of the museum is Dr. Samuel West, licensed psychologist, PhD in Organizational Psychology. Below are some excerpts from their website, www.museum of failure com.

> **Innovation needs failure.**
> The Museum of Failure is a collection of failed products and services from around the world. The majority of all innovation projects fail, and the museum showcases these failures to provide visitors a fascinating learning experience. Every item provides unique insight into the risky business of innovation. Innovation and progress require an acceptance of failure. The museum aims to stimulate productive discussion about failure and inspire us to take meaningful risks.
>
> Making mistakes is a fact of life—and that's not a bad thing.
>
> The Museum of Failure brings together over 159 products and services that were a total flop but also paved the way for other great inventions. Failure is the mother of success, after all!
>
> **Learning From Their Mistakes**
> Embracing failure and taking meaningful risks allows for real innovation and progress. With a unique insight into the risky business of innovation, the museum aims to inspire

and stimulate productive discussion about learning through our blunders. Everyone falls, but what's most important is knowing how to get back up!

"Do not dwell in the past, do not dream of the future,
concentrate the mind on the present moment"
—Gautama Buddha

Summary

Unintended or unwanted outcomes, whether they are called errors, mistakes, failures, etc. are inevitable. They are endemic to any complex biological organism, including us humans. Given the novelty of each situation we encounter and the complexity of our brain and body, total consistency is not in the cards. We are not machines; even machines have glitches. The best we can do is to accept the unintended outcomes as opportunities for growth and improvement. That is easier said than done, but self-recrimination is not particularly helpful or productive for future learning and performance.

Your Notes

CHAPTER SEVEN

Learn by Drill or Play?

It depends!

"No one is born with skill. It is developed through exercise, through a blend of learning and reflection that is painstaking and rewarding. And it takes time."
—Twyla Thorpe

The foundation for learning and applying anything including golf is the development of problem-solving skills. When given problems to solve, learning is being supported more effectively than doing drills or following directions.

Hebron

As a coach, one of the many things I now do differently than in the past is I suggest more play than repetitive drills. Make prep time play time. During golf instruction and training, the aim is developing skills for a game that is played in ever-changing environments. Develop flexible and portable skills. Develop the mindset of adapting.

Play games when training. For example, during training I have students, both professional and club golfers, change targets every swing, as required when playing. Also change clubs, size of swings, and use different swing speeds.

When using the same target, create high, low, long, short, left, and right ball flights to that target. Play golf to learn golf. On the green, putt and roll the ball long, then short, then left, then right of the hole, then to the hole. Play putting games.

PLAY = **P**owerful **L**earning **A**bout **Y**ourself

PLAYFUL = **P**owerful **L**earning **A**bout **Y**ourself **F**inds **U**seful **L**earning.

Hebron

For years, many people believed that the fastest, most productive way to learn tasks was to repeat those over and over, using drills and expert models. Research by Dr. Langer of Harvard, and many other leaders in education now suspect there are some problems with this long-held approach. Doing something over and over may support progress during early training, but random training supports long-term learning and the retrieval of skills beyond short-term performance. When training your golf swing, develop random outcomes, high, low, left, right ball flights.

The approach of repeating drills does not leave much room for self-discovery and rethinking information. There are more effective ways of mastering skills. Dr. Langer calls it "mindful learning," as opposed to mindless, and points out long-term learning requires much more than drills founded on memorization, repetition, and rote. Conventional drills depend on automatic behavior, which stifles creativity and undermines self-esteem as we struggle and often fail to learn techniques adequately with a mindless drill approach.

Using drills leads to boring, mindless, non-focused training sessions that often leave people with little or no personal insight into the skill they are trying to learn. Using a repeating unthinking manner to learn basics can almost guarantee mediocrity. Drills are normally based on a *how* to do the task, and do not give individuals the opportunity to gain insights and options they did not have in the past.

Expert golfers become experts by developing various ways to use the very basics some golfers are trying to improve with mindless repetitions. Learning happens best when skills are performed when conditions are not static. Golfers are never faced with the exact same shot or conditions, with experts adjusting their basic skills for every different context. Drills are more useful after a skill is learned than during learning.

The aim is to have a flexible and portable swing; if it were consistent, you could only hit one kind of shot. Simply memorizing the execution of drills will not lead to deep learning, whereas awareness and self-directed learning can. After a skill is learned, repetition drills are more useful.

Using a general just-in-the-ballpark concept of a task and moving away from an obsession of trying to get it right in drills can bring on the joy of learning. When learners stop judging themselves when doing drills and become more involved in the whole experience of learning through awareness, pleasure and productivity become partners on the path to long-term progress. I know that asking golfers to move away from using drills when learning is controversial, but it is a decision that cognitive science and my experience show supports deep learning.

"All learning is Play."
—Albert Einstein

The Value of Play

- We play to learn; in life we did not have to learn to play.

- Play never tries to memorize—but has wonderful long—term memory.

- Play never has failure, only yes and no results that are never wrong.

- Play does not try to be exact, but often is.

- Play does not rely on exact directions, but always gets to its destination.

- Play tries what's new and different, and with little fear.

- Play takes risks but is never anxious.

- Play has many Ah! Moments that last forever.

- Play happens in safe environments. Learning requires a safe environment where outcomes are not being corrected by others.

A significant component of the early sport experience of current elite athletes was their widespread involvement in a range of both deliberate free-play and organized sports. Whether it is back-lot stick ball, driveway basketball, keep away (i.e., informal rugby), deliberate play is normally regulated by flexible rules, adapted from standardized sports rules, and are set up by the participants involved in the activity.

Deliberate free-play activities promote interest over focusing on trying to make sports fun. When play is interesting to individuals they stay involved, even during their unwanted outcomes. When involved with free-play, there is less concern with the exact outcome of motion than with the enjoyment of the behavior. Deliberate play in sports can have immediate value in terms of motivation to stay involved and it also has benefits related to the ability to process information in various sporting situations. Obtaining a skill in each different sport will transfer to and influence other sports skills.

Motivation based on self-regulation supports the idea that early intrinsically motivating behaviors (deliberate play) have a positive effect on a) staying motivated, b) becoming more self-determined, and c) being committed in future sport participation. A variety of studies from numerous sports shows that deliberate play is superior to deliberate practice in the development of athletic skill for recreational as well as elite athletes for example, (Ryan & Deci, 2000; Berry & Abernethy, 2008). From a skill acquisition perspective, deliberate play serves as a way for athletes to explore their physical capacities in various contexts. This was found to be true for elite professional hockey players who spent more time in deliberate play than deliberate practice activities before the age of 20.

Elite baseball players were also involved in more deliberate play than those who stayed recreational players from ages 6 to 12.

Expert game-decision makers in Australian-rules football invested a significantly greater time in varied deliberate free-play activities while playing basketball, football, hockey, within a space of two years than non-expert, game-decision makers. Deliberate free-play in various contexts then appeared to provide a broad foundation of skills that helped to overcome the physical and cognitive challenges of various sports as well as their main sport. Still, excellence at an elite level in one sport does not guarantee excellence in another sport, for example: Michael Jordan—from basketball to baseball; Tim Tebow—from football to baseball.

A basic law of human development; people can be nervous about change. But people are normally not nervous about play, which supports learning and reaching our optimal potential through child-like play. My course, Smithtown Landing, has a very good 9-hole Par 3 course called THE LEARNING LINKS, WHERE WE PLAY TO LEARN.

Playing-to-learn environments speak loudly to an adult's child-like instincts of self-discovery play. Playful approaches to progress reproduce what was emotionally powerful to a child, for adults.

When playfully training with insights about *what* to do with the golf club, and not trying to copy how-to directions from a perceived expert for moving the body, progress emerges. By seeing both workable and unworkable outcomes as equally usable feedback for future reference not as good and bad, (success and failure) helps progress emerge.

General concepts that are just in-the-ballpark, and not technically perfect, are very valuable tools for making progress. "Just shoot the ball up," or "just swing the bat," or "just swing a golf club," without following specific details about how to do it, is the suggestion here.

The natural environment puts forward the most useful model for what to do. The weather dictates how we dress. Business responds to the needs of the customer. A doctor does not decide what to do (how best to use the core information gained in medical school) until they see the patient. A lawyer cannot use what they know about the law until they know what kind of case they are working on. To be fully informed, be aware of the environment, it is a true catalyst of creativity. The golf course provides insights for what to do with the golf club.

Yazulla

The foundation of any learning situation seems to be *play* in the form of satisfying some curiosity: a toddler scrawling a crayon over a piece of paper, a child bouncing a ball on an uneven surface, or a teenager doodling while daydreaming while listening to music. The toddler is learning eye-hand coordination, critical for any mechanical activity. The child is learning that a bouncing ball does not always bounce straight up. The daydreaming teenager may be pondering the solution to some problem, a plan of action or just wishful thinking. In any case, there is some activity, apparently random, but with a goal in mind. These loosely directed behaviors eventually lead to more organized behaviors, the proficiency of which requires practice and drilling, or in other words, discipline. The question really is, once a level of proficiency is achieved, what is the best way to increase that level of proficiency and then maintain it?

The common method seems to be disciplined practice and drilling. To drill and to cram have much in common as strategies when learning any new task. The purpose is to set up and reinforce the brain circuitry for a fundamental required for the task. I think this is true for academia, sports, trades, the arts, etc. These fundamentals form the foundation upon which one can afford the luxury of playing, diversifying, and experimenting to expand one's skill. The distinction to make here is that there are endeavors/professions that require disciplined instruction at the beginning and those that do not.

For thousands of years, people did not just jump into a profession. They were sent out as apprentices, and only after years of work with a mentor, did they become a master. The same is true for some professions today for obvious reasons, i.e., surgeons, high-voltage workers, jet pilots, glass blowers, crane operators, and so on. Other professions can evolve from earlier interests, hobbies, or play. These include any sport, history, literature, astronomy, archeology, biology, performing arts, etc.

What about Drills? As examples, there were two very important lessons about the value of Drills for me. First, in the second grade we all learned the 12-times table during the school year. It was sheer rote and drilling, day after day, until those values were memorized and recited on-demand. Knowing the 12-times table became as automatic as reading. Secondly, with Biochemistry in college, there seemed to be no rationale to memorize the structure of the vital amino acids. You simply learned to identify them by their structures and their formula by sight. However, the entire field of protein chemistry and proteomics depends on the detailed knowledge of amino acids.

Once these basics were committed to memory, then there was the opportunity to play and diversify. For math, the effect on me was that I could think in numbers, since I understood that multiplication and division were shortcuts for addition and subtraction of number columns. I could estimate squares and square roots, as well as products and division of larger numbers just by relating them to the 144 numbers in that "12 by 12" matrix. I still play and do all sorts of number games in my head simply because of the intense drilling of basics at the age of seven in the second grade.

As for Biochemistry, others (not me) use the basics of amino-acid structure to determine the structure and function of proteins in any environment (proteomics), leading to an understanding of diseases. This is because proteins are major components and targets of pathological organisms. In September of 2021, one strategy used to develop vaccines depended on identifying vulnerable spike proteins on the virus that causes COVID-19. If it wasn't so serious, one could consider this vaccine search as play in that it involves much trial-and-error, hit-and-miss experiments based on educated estimates as to how proteins interact with each other. However, with detailed knowledge of amino acids, the 3-dimensional structure of proteins in any environment could be calculated and used to determine vulnerability to potential antibodies.

An analogy with all endeavors is learning how to write. First, you learn letters, to form words, then sentences. However, after learning the basic letters of the written language, the formation of words and then sentences follows certain rules of spelling and grammar. The goal of a sentence is to convey the thought of the writer to a reader. The random placement of letters in a so-called sentence would fail this purpose. "od oyu tge ti" makes no sense but rearranging the letters into the words "do you get it" now makes sense. These words can be rearranged and still make sense. For example, "get it do you" as in "Yoda speak."

For our purpose, in sports and academia, going through the sequence of learning steps from simple to complex carries little or no risk. The consequence of this is that once fundamentals are learned, the focus can be on play and diversification in a safe supportive learning environment. More on this is in the section on Practical Suggestions (Chapter Ten).

Hands-On or Hands-Off Coaching? *It depends.*

Yazulla

Two alternative coaching techniques involve the use of either a hands-on approach or hands-off approach. When learning a golf or tennis swing, an instructor or machine may try to physically guide the student through the desired motions. In this hands-on approach, the limbs and body of the student are passive, allowing the instructor or machine to guide the movement. Alternatively, in a hands-off approach, the student actively controls their own movement, with the instructor providing verbal or visual guidance. The question is how useful is the student passive, instructor hands-on approach, compared to the student active, instructor hands-off approach?

To address this question requires some very general background. As mentioned in Chapter Four, our brain's control of movement depends on sensory feedback from sense receptors in the tendons, muscles, and joints. Joint receptors provide information about the angle of our joints and positions of our limbs; tendon receptors measure the stresses and forces placed on the joint by muscles as they contract; muscle receptors measure the degree of stretch or length of the muscle, not its force (a contracting muscle gets shorter). The information, supplied by these receptors to the brain, is called 'proprioception' and 'kinesthesia.' In addition, inputs from the visual system, as well as from our sense of balance in the inner ear and eye movements contribute to our sense of position sense. So, what does this have to do with hands-on and hands-off coaching?

If an instructor or machine moves your arms (hands-on) to take you through a golf swing, you follow passively but you are not actively contracting your muscles. Your muscles stay mostly relaxed. In fact, your muscles may actually oppose the movement. You are aware of the movement because sensory information from your joints (angle) and muscles (stretch) tells you where your body parts were at the start, where they are now, and how they got there. However, this is only partial information because the passive movement reduces the sensory feedback from the tendon receptors. These tendon receptors respond to the load placed on the limbs by how much force is applied to the bones by the contracting muscles. With passive movement, you have experienced the path the body parts took during the swing, but not the effort the muscles would have exerted on the joints to move those body parts. The difference is feeling the movement, rather than doing the movement.

The topic of machine-aided learning includes Haptic Guidance (haptic, Gk for touch), meaning anything related to the sense of touch as tactile feedback to assist motor learning. Detailed discussion of haptic guidance, robotics and biomechanics are beyond the scope of this volume. Haptic Guidance is used with some success in clinical settings of physical therapy following trauma, surgery, stroke, and other disease states Hobbs & Artemiadis, (2020, for review). Along with massage, the passive movement of limbs stimulates muscle and joint receptors, may activate spinal reflexes that stimulate opposing muscles, and reduces joint lockup by keeping them mobile. In the case of an intact spinal cord, passive movement will stimulate the sensory motor cortex and associated intracortical circuits that could participate in the recovery of movement following injury.

There is a very large literature that addresses the topic of passive training in controlled laboratory settings. Robotic devices are used mostly to passively move the hand and arm in some trajectory (i.e., dart throwing, drawing, or moving a joystick). The use of external, robotic devices ensures the reliability of the guided repetitive movements. The general finding is that passive training (hands-on) can temporarily enhance motor learning of simple tasks see for example, (Reinkensmeyer & Patton, 2009; Beets et al., 2012; Wong et al., 2012; Bouchard et al., 2015; Chiyohara, et al., 2020).

To quote from Reinkensmeyer and Patton, (2009), relating to learning simple tasks "Indiscriminate application of this approach (passive training), however, does not seem to produce much benefit in the amount of learning compared with visual presentation of the desired trajectory.

Thus, visuomotor learning mechanisms do not appear to be much enhanced by the addition of haptic information."

Contrast the above with Chiyohara et al. (2020) referring to learning an arm motion, then using that motion to throw darts "In conclusion, the present study indicates that passive training improves proprioceptive acuity around the trained joint and learning performance of another type of movement involving the same joint."

Reinkensmeyer and Patton, (2009) do not dismiss passive learning. They note its value for tasks that are too difficult or too dangerous to learn, for example, race car driving, diamond cutting, vascular surgery. The passive training guided and constrained would allow a person to try something new and to experience the dynamics of the movement while constraining errors in new, difficult, or dangerous tasks.

Passive training prescriptive-proprioceptive training, haptic guidance is also being used for motor learning in sports and other every-day activities. Some coaches use passive training hands-on during instruction and believe it to be beneficial for the student. Obviously, a student has to demonstrate the learning hands-off by actively performing the task: swinging the club, bat, throwing the dart, using a pool cue, hand position and movement for throwing clay on a wheel, holding a paintbrush, and so on. Once the guided movement is learned, how long does that learning persist? Do students fall back into their old habit? Is there a learning curve in which the learning rate of the desired movement improves in the future? How well does the passively learned movement translate to the golf course, for example. Without controlled laboratory studies, it falls to instructor/student experience to answer these questions and determine how to proceed.

From my perspective, I see passive training hands-on as a bit of a shortcut, to get one's foot in the door. Some people may need a bit of help with duplicating a movement or position, for example, how to place their fingers, and move their wrist to throw a curveball as opposed to a slider, varying ways to grip a golf club, or simply how to maneuver chopsticks.

In the final analysis, it is the multisensory input and feedback from the skin, muscles and joints following active motion that are required for coordinated motor movement as coded in the brain.

Hebron

Movement is encoded in the brain *NOT* in the muscles. Physically guiding students' motions and the use of aids do not send the same feel and other information to our CNS and our brain that movement without aids does. I prefer a method of coaching that minimizes the use of such aids.

Here is just a little background information on my rationale. Over the years, many scientific studies have been done at research centers and universities on the topic of motor learning. I have referenced many of these numerous times.

For example, Harvard's Connecting Mind-Brain to Education Institute, UCLA's Learning and Forgetting Lab, Vanderbilt's Peabody School of Education, The University of Washington, The Human Brain Project, National Institutes of Health, Allen Institute, Salk Institute, and many other universities and research centers have invested and continue to invest millions of dollars to inform educators about the human brain's connection to learning. I have been learning from these resources for years and they can be taken advantage of by others.

The following example is from Dr. Fran Pirozzolo, Ph.D. Neuropsychologist Director of Mental Training at Northwestern University, on giving Feedback and Physically Guiding Learners. He has been coach to several professional sports teams, PGA tour players, Navy seals, and an excellent low handicap golfer.

Dr. Fran Pirozzolo from the video, *Challenging Practice Leads to Effective Learning*, talks about how traditional methods of practice are not optimal for learning, and discusses alternative methods that are much more effective. Practicing in distraction-free environments and performing one skill over and over might feel easy, but it leads to worse learning than when athletes are encouraged to overcome challenges during practice. Watch this video to find out more! http://www.lastinglearning.com/2015/10/25/challenging-practice-leads-to-effective-learning/. Other videos are available.

- ○ Study results challenge the assumption that all feedback and physically guiding will improve later retention of knowledge and skills.

- ○ Empirical evidence shows that delaying, reducing, and summarizing feedback is best for long term learning to occur!

- ○ Frequent & immediate feedback can degrade learning!

- ○ Feedback after every trial boosts short term performance but it impairs later learning.

- ○ Feedback needs to be delayed to be best.

- ○ Also consistent is the guidance hypothesis. Providing aids and physically guiding performance makes players depend upon the teacher propping up performance. This is a key to performing under pressure—all the immediate feedback that was available in a lesson is no longer available.

- ○ Crutches that are removed later and are no longer present impair learning and retention.

Summary

Is it better to Drill or Play? Of course, it depends on the activity and the stage of competence you are at. Both are important. Regardless of whether you are engaged in sports, academia, trades, crafts, medicine, law, business, etc., initial interest may have been to satisfy curiosity by play. Competence then requires serious application of practice, training, drilling, whatever you may call it. Once a level of proficiency is achieved, people often tinker, play, experiment; "what if I change this?" Innovation often happens when experts tire of the status quo and try something different—build a better mousetrap. So, the cycle goes on: play, drill, play, and so on with progress. Circumstance and the individual determine what is better at any time during skill acquisition. Presumably, the goal is to learn, perform, improve and, at the least, have some enjoyment and satisfaction in the process.

Your Notes

CHAPTER EIGHT

How Useful Are Expert Models?

It depends.

Hebron

EXPERT MODELS It is not uncommon for a student to spend time trying to copy an expert model. My question is how did that expert learn what they do? For me, the value of watching an expert is they can get others excited about doing what the expert is involved in. But trying to do exactly what an expert does is not as useful as a personal LEARNING MODEL. Why? A learning model is a description of the mental and physical mechanisms that are involved in the acquisition of new skills and knowledge and how to engage those mechanisms to encourage and facilitate learning.

Studies by Dr. Willingham at the University of Virginia point out there are **NOT** 3 types of learners. The idea that some people learned better by listening (auditory), doing (kinesthetic), or seeing (visual) something done, has been debunked. At best these are only preferences for learning. Controlled studies showed no difference in retention when people were taught in versus out of their preferred learning style. All senses are involved simultaneously when doing anything including learning. (Willingham et al., 2015, for review; Rogowsky et al., 2015, 2020).

Students, both professional and club golfers, come in different sizes, ages, talent levels, past experiences, amount of time to train, and motivation. **Every golfer is a novel and unique individual and also is the look of every expert in motion while doing the same thing**. Ask professional golfers how best to create a hook-ball flight, you may hear a different way from every player. Ask a professional baseball pitcher how best to throw a curveball, again you may hear a different way from every pitcher.

When training to improve the outcome of a student's driver swing, for example, the information being used to support a change can be different for each student. There is no attempt to fix anything with a learning model. The aim is simply to improve the student's outcome. Skill growth is a personal self-development, useful than strong focus not an attempt to look perfect or fix anything.

Expert Models

Hebron

An expert model was thought to be a good representation of what to do, and also good for changing unworkable habits. IT IS NOT. The idea is that the more often learners see an expert model; the stronger the mental blueprint of the model would be, hopefully translating into increased ability to perform and change unworkable motions. IT DOES NOT. Coaches from many sports have supported the use of expert models for years.

I do not have a specific model. There are some basic components of a learning-developing approach to instruction available that are very different from a teaching/fixing to get something right method I once used. My view now regarding golf instruction is supported by studies from Professor Gibson F. Dardon of Radford University Dardon, (1997) and many others in the field of motor learning. Their studies indicate that the use of only expert models encourages imitation, not deep learning. Much of the following information is based on my experience and their insights.

Expert models often do not guide learning into retention of the skills needed to perform in changing conditions and situations. Typically, expert models provide demonstrations of motion that we expect people will be able to copy. In fact, there are many popular video instruction systems that compare expert models to a learner's motion. This form of training leads to a performance that resembles a simple reflexive action, not deep learning.

Because of the various conditions that playing golf can present, golfers would be better served if they developed some creative skills so they could learn to make adjustments. Developing a conditioned reflex using drills to copy expert models has been found not to be the most useful learning strategy for learning to play golf, or any sport for that matter. On the other hand, guided trial-and-error adjustments support long-term learning.

Variable training i.e., different speeds, swing sizes, paths, and clubface alignments in a golf-swing motion promotes more effective learning of the student's own rules and feels of an effective motion. These variable motions help people learn the cause and effect of motion patterns. This variety of motions during a training session can result in a stronger memory of the basic skills needed to influence the club head, and shaft through impact.

Consider pitching a baseball, the basic movement of the arm is rather stereotyped. However, each pitcher has idiosyncrasies in their delivery that they develop through trial-and-error training. The same is true for a golf swing. Increasing the variety of motions during a training session results in a stronger memory of the basic skills needed to influence the club motion through impact. This influence is necessary depending on the changing circumstances one encounters on a golf course. As many recreational golfers have learned, repetitive practice on the perfect lie of a mat at the driving range, often does not translate to success on the course.

Trying to repeat an error free performance (i.e., standing in some swing ring, swinging over and over) imitating an expert model appears to ignore what we now know about learning motions. Errorless practice and repetition are poor deep-learning strategies. After a skill has been learned, repeating it over and over can have some value to highly skilled performers, but not for learners. There is real value in missing a foul shot, or missing a serve in tennis, or missing a shot in golf. Players can learn from their unintended outcomes.

> *Failure is just an opportunity to start again, but just more wisely*
> —Henry Ford

Workable and unworkable actions, (both mental and physical) must be experienced for deep learning to take hold. The exclusive use of expert models and drills sidesteps opportunities to experience a wide range of motions. The brain thrives on a variability of experience. Becoming aware of the differences in workable and unworkable motions is an important stage of long-term retention of any skill. Introduce some poor outcomes, hooks, and slices when training is the suggestion here.

Golfers looking for long-term retention of skills should avoid learning a task under static conditions. Real life performance conditions require problem-solving skills and adjustment strategies. If golfers train and practice on a range, I suggest make believe you are playing holes, or change clubs after each swing, as you change targets and size of swings. I have found this kind of variation in training even for new golfers more useful for long-term learning than imitating expert models using a how-to-do-list.

Golfers who educate themselves about that process of learning a skill and not look for immediate success that often leaves as quickly as it arrives experience deep learning. Learning with a mindset of finding options that lead to increased performance during training is the suggestion here. However, since skill development at first is an internal process and therefore invisible, evidence of a learning transfer may show up at any time during training or play. Learning takes time.

Students who gain insight into the difference in temporary effects of instruction and long-term retention of skills, or the difference in trying to fix a habit and developing a skill, growth tends to follow. Using guided self-discovery and problem-solving approaches opens a path to long-term retention that drills of how-to-lists and expert models do not provide.

Yazulla

It seems to me that the success of Expert Models depends on mimicry, the ability of some animals, including humans, to replicate movements they observe as performed by others—as in "monkey see; monkey do." Youngsters learn early by copying behaviors of their parents, siblings, and friends— "watch me, see how I do this." Instruction manuals and videos exploit this ability to mimic and follow directions. The importance of providing such information when confronting the assembly or operation of any complex mechanical or electronic device cannot be overstated. As they say, "a picture is worth a thousand words." However, no manual or video can do the job for you.

My experience with expert models is positive, with a caveat. Everything goes well until you have to replicate the fine motor movements of the activity you are trying to learn. As an example, I offer my recent experience with masonry work. Not knowing much about concrete, I found the variety of mixes available at box stores to be a bit overwhelming. With so much how-to information available on YouTube™ from playing a guitar to rebuilding a car engine, I thought I would look up masonry work as a part of a new project. I came upon a site by Mike Haduck, an old timer with a straight-forward presentation of information about all sorts of topics on masonry. With his mantra of "It's no big deal," it is a pleasure to watch an expert clearly describe what he does and why.

After watching many episodes and repeating, I felt ready to attempt a repair of cinder block steps. I chose an all-purpose, quick-set concrete mix that was suitable for thin or thick application, and applicable for horizontal and vertical surfaces, as found with steps. Mixing the concrete with water to the proper consistency was easy. I just followed Mike Haduck's directions and observed the consistency change as I added varying amounts of concrete and water and mixed it. It was straightforward to achieve the proper consistency. So far, following the expert model worked like a champ; just use your eyes and follow the recipe, like making pancakes.

Mike Haduck's other mantra was "keep it wet to make it stick." The surface had to be kept wet. Now the hitch, applying the concrete was not so easy. His fluid motion of putting concrete onto the trowel and, either with a backhand or forehand, applying it onto a wall looked easy, but it was not. There was the issue of concrete consistency and surface wetness. The vertical surface of the step was too wet, and the concrete just slipped off onto the flat surface. If the surface was too dry, the concrete peeled off the surface. It took many trials to get the proper mixture and moisture level of the surface and feel of the motion to apply the concrete. In addition, this was quick-set concrete;

so, this is like a speed test, a limited time to get through the amount of prepared concrete. Only mix as much as you can use in fifteen minutes. This takes practice too.

The value for me about an Expert Model is that it got my foot in the door. It got me over the inertia to try something brand new. I felt some confidence when I started. It worked for me when conceptual issues were in play, as in, following a recipe to mix the concrete. But when the sequence involved complex motor movements, the expert model gave me the constraints of the arm and wrist movements (sideways, not overhand), but reducing the margin of error within those constraints required practice. By constraints of the movement, I mean the limits in which the movement must occur horizontal for a baseball swing, as opposed to vertically pounding a stake into the ground. The expert model got me out of the gate, but refinement of the movement to the finish line took practice.

Alternative—Learning Models

Hebron

Often when expert models are used, a student's self-confidence can be damaged. Without some level of confidence, progress for anyone would be difficult. The use of a learning model, in which people will be encouraged to take into consideration the strengths and limitations of an individual learner, is the suggestion here.

Any pressure to do it like an expert can prevent people from discovering what works best for them. Trying to copy an expert model can stop people from exploring different feelings for timing for example, and overlook the power of self-awareness, and self-assessment, which are more useful for learning than trying to imitate an expert model. An expert model may get one's attention and may even increase motivation, but with respect to learning or thinking about the skill, expert models often fall short when compared to general descriptions and self-discovery found in a learning model.

A learning model hypothesis suggests that the most effective approach to learning is the one that is only slightly above the learner's current skill level, promoting persistence and self-esteem during learning (McCullagh & Caird, 1990).

A learning model suggests that golfers would be better off if their personal perspective was more or less just in the ballpark, and not necessarily a perfect picture. Using a non-specific learning model can be a more effective method for allowing people to become skillful at their current level, especially when combined with accurate feedback. Several studies question the need to give people detailed explicit information on how to execute motor skills (Rink, 1994).

People only require general information about appropriate movement patterns to reach the goal of very accurate or explicit patterns of motion. I and others have found observing a learning model (a non-specific generalization) is more effective than using an expert model. Allowing people to explore possible solutions is a critical aspect of the learning process. From this view, expert models may indeed restrict learning.

It may be difficult for many of us who have used expert models and drills, including myself, in the past to believe that by replacing the expert model with a learning model will enhance learning. But I can say from experience it does. I suggest working with general information without the use of repeating drills rather than using very specific details. Based on what is now known about the nature of learning, golfers and instructors should consider moving away from using drills and expert models and use a learning model that supports flexibility and experimentation.

Yazulla

Back to the topic of cramming that involves intense, focused activity just prior to a performance. In academia, cramming is common the night before, or the morning of an exam. Often the student has not kept up with the material since the previous exam. The rationale is that one can absorb a large amount of new material in a short period of time. Unless you have photographic memory, most of the information will not be retained because of the limited amount of time for the memories to consolidate into longer-term memory. The hope is that enough of it will stick and be represented in some of the questions on the exam. In my own experience and with numerous students, cramming offers some partial success, perhaps worth a question or two per exam but not enough to optimize academic performance.

As a strategy (discussed in (Chapter Five) the worst type of cramming is an all-nighter. As a faculty Professor, I advised very strongly against all-nighters because it is a very inefficient learning strategy. Mental and physical efficiency is disrupted by the lack of sleep from an all-nighter, and this effect may last for days. Just consider the effect of jet-lag. To adjust, it takes roughly one day per hour shift of time zone. An all-nighter is the equivalent of flying from New York to Tokyo while staying awake. I would advise students that if they could not get the information in by 10 pm, cash it in, and get a good night sleep. At least they would be fresh for the exam and could not blame a poor grade on sleep deprivation. The other excuse was sleeping-in because they finally fell asleep at 5 AM and did not or could not wake up in time.

Cramming negates two important factors in learning: time gaps between learning sessions and sleep. Both of these are needed to consolidate memories from short to long term. Synaptic connections among brain nerve cells need time to be strengthened to form long term memory circuits. Rest between sessions and particularly deep sleep facilitate this process.

Cramming prior to an exam or performance should be limited to a superficial overview, in order to trigger recognition and recall of previously learned material. Any form of new learning and its retention in the short term is unlikely with cramming. I am sure there are exceptions; I have known students who crammed with the purpose of just getting a C. Their goal was not to get an education, but simply to pass the course with as little effort as possible. In my opinion, cramming is not a successful long-term strategy in any endeavor in which proper preparation is required. I have found that cramming is an inefficient way to learn and is likely to be detrimental to any impending performance. The fact that you are cramming is an admission that you are not prepared. This realization and expectation of a negative outcome in itself will increase stress, further suppressing performance.

Up until now we have been discussing individual learning and performance. However, when dealing with teamwork, drilling takes on more significance. It is one thing for an individual to train to the beat of their own drum, and quite different for many to train to the beat of the same drum. When a performance depends on the coordinated behavior of each individual, drilling and training becomes essential. Consider the complex performance of a marching band, drum–and–bugle corps, kick line, plays in football, choruses, orchestras, pairs skating, dancing, theater etc. Individuals, regardless of how proficient they are individually, could not come together for a seamless performance without having drilled/practiced extensively as a group. Even small groups, performing expertly, are said to work as a well-oiled-machine, all parts of one entity, working together.

Summary

Expert models certainly serve a purpose, perhaps at the early stage of learning. They may look like the ideal, but they should only provide a framework within which your own behaviors can be modified. No one can exactly repeat their own movements every time, much less drilling to achieve that of an expert model. Remember, the style of an expert model is usually based on a single accomplished individual, who is NOT you. So, make use of the model, but adapt it to your own style.

Your Notes

CHAPTER NINE

Narrow Focus or Wide Attention; Diversify or Specialize?

It depends.

"The mind has two abilities—one is to focus, the other is to expand and relax."
—Gurudev Sri Ravi Shankar

Hebron

Attention

I found the following in *The Oxford Handbook of Thinking and Reasoning*, edited by Keith J Holyoak and Robert J Morrison to be very interesting (P. 767).

"Our ability to strongly focus can lead to a narrowness of insights and vision. This can actually stand in the way of more effective ways of thinking, locking us into rigid patterns."

I like to tell the story of the driver of a car who drives through a stop sign during a heavy rainstorm. When stopped by the police he said, "I am sorry officer, I was focused so hard on the rain, I did not see the stop sign." Suggestion; use wide attention. Also, we often hear, "I was focusing so much on the line of the putt I forgot to hit it." Wide attention is often more useful than strong focus.

Narrow Focus

Yazulla

To focus during a learning situation, means the ability to shut out distracting stimuli when attention is devoted to the task at hand. Humans cannot give full attention to learning or attending to two sources of information at the same time. Multi-tasking refers to performance rather than learning, and even with performance, multitasking is overrated. Examples are an attempt to eavesdrop on two conversations or trying to read while someone is talking to you. What you are really doing is time-sharing, alternating your attention from one conversation to another. You may pick up snippets of both and hope to fill in the blanks later.

The same thing happens if someone is talking to you while you are trying to read. Often in this case you will have little idea of what you read or what was said to you; both will likely need to be repeated separately. You stop, look at the person and say, "What?" Then afterward, return to your reading and reread the passage that was interrupted. The point is that the human brain has great difficulty processing simultaneous information coming in from the same sense i.e., hearing, or different information from different senses i.e., touch and sight. This differs from watching a movie or theater performance, for example, in which all senses are directed at the same source.

The competing sensory information simply is not processed efficiently in memory. This topic of selective attention was addressed when discussing the problem students have taking notes while the teacher is lecturing. It is very difficult to listen and write at the same time. However, stenographers' shorthand, and court reporter's stenotype machine are very proficient at doing just this, with much practice and training. However, I have been informed that the goal is to transcribe what is heard, not retain the content of the information.

How long should anyone focus on any given learning task? Well, that depends on numerous factors, particularly age and one's attention span, and level of interest in the topic. In general, attention span increases as we get older, and of course, differs among individuals. The attention span of children may be as little as 5 minutes, and it gets longer as we enter adulthood, then it decreases again. In high school and college, 40–50-minute daily classes are the standard and seem to be a good time limit for the student and teacher. However, classes at 75–90 minutes twice a week, and at 2½–3 hours summer school and night school stretch and greatly extend the resources of attention for all concerned.

The longer classes are tough on the voice of the teacher, and physically and mentally demanding on the mind and body of the students. In my experience, these classes are either cut short or broken up with coffee-breaks. As a result, less information is covered and often more superficially than in the shorter classes, simply because of mental overload and fatigue. When should you end a learning session? When your mind wanders, when your eyes glass over, when you are tired, there will be a diminishing return on time spent and topic learned. It is time to stop and relax.

Hebron

I have found that long training sessions with new information are not as effective as breaking a long strong-focused session into several short sessions with a short break in between. Once boredom sets in, periodic breaks help to consolidate the newly introduced motions. Change the context of your training every 5–10 minutes. For example, when training golf club alignment, spend 5 minutes on tee, 5 minutes on green, 5 minutes on fairway, 5 minutes in the bunker, rather than 20 minutes in one place.

This brings up sustaining "**AWARENESS**"

Alert **W**ith **A**ll **R**esources **E**ngaged **N**aturally **S**ensing **S**urroundings.

Awareness is not something we get or develop; it is something we already have. Awareness is "allowing." Attempting to focus hard for a long time often brings on overthinking about results and hinders awareness during learning and performing. This outward hard focus on results is brought on by a culture that is concerned with spending lots of time getting it right, which can prevent awareness in long training sessions in the same context.

Awareness recognizes practical non-specific opportunities for creativity and imagination. This begins with exploring the power of being mentally quiet in short sessions of training in different contexts and what that actually feels like while being in the present.

Many assume that increased focus is always better, but as Lehrer has pointed out, they overlook that intense focus comes with real tradeoffs. When individuals are not filtering out the world by trying hard to focus, they end up subconsciously letting in useful information. The brain subconsciously and naturally considers all sorts of analogies, which provide useful insights for learning and solving problems during non-focusing. When your brain is supposedly doing nothing (such as not focusing) it is doing a tremendous amount. I call this "wide attention."

"Creativity is the result of time wasted."
—Albert Einstein

Dr. Marcus Raichle, a Distinguished Professor at Washington University is a pioneer in the development and use of positron emission tomography (PET) and functional magnetic resonance imaging (fMRI) to study brain function. He and colleagues demonstrated that during non-focusing, or what he calls a default stage, there was elaborate electrical conversation going on between the frontal and parietal parts of the brain. Why was the brain so active during non-focusing or daydreaming? He found that the prefrontal cortex was falling in sync with brain areas that normally do not interact directly. It is when we start to daydream that these areas of the brain begin to work closely with each other, making subconscious associations by connecting new experiences with prior experiences (Raichle, 2015, for review).

In our wide-attention stage, the brain blends different kinds of skills and concepts that are already encoded throughout the brain. During wide-attention, instead of just responding completely to the outside world, the brain will start to explore its inner database searching for relationships in a more relaxed fashion. This is often when "Ah!" moments arrive. This relaxed mental process often runs parallel with increased activity in the brain's less linear and more creative right hemisphere. Suggestions: stop trying so hard to focus, and let insights just arrive from your subconscious wide-attention mind.

Strong outward attention can actually prevent us from making the connections in the brain that lead to workable insights. During wide-attention, the brain uses conceptual blending. This is the ability to make separate concepts and ideas co-exist. Breakthroughs often arrive when old ideas or past solutions are applied to new situations. Instead of keeping concepts separate, the brain will subconsciously blend them together when we are not trying to focus.

Wide attention brings forward information from our subconscious mind that supports learning, performing and creativity. For example, the corporate history of innovation at 3M labs is actually based on their scientists' taking breaks from thinking about ideas for new products. Today Minnesota Mining and Manufacturing Company, called 3M, sells more than fifty-five thousand products, nearly one product for every employee. The essential feature of 3M innovation according to Larry Wendey, a vice president in charge of corporate research, is the flexible attention policy that I would call brilliant. Instead of insisting on constant concentrations there is a 15% rule. Every researcher is allowed to spend 15% of their day daydreaming, allowing speculative insights to surface. The only requirement is that the researchers must share their ideas with their colleagues. This sharing is much like how the brain blends different concepts together. Scientific study of insight supports 3M's flexible attention policy.

People who score high on a standard measure of happiness solve about 25% more insight puzzles than people who are upset or feeling frustrated. The relaxed feelings of delight can lead to dramatic increases in creativity and learning. Mental relaxation makes it easier to daydream and pay attention to subconscious insights. People who consistently engage in more daydreaming score significantly higher on measures of creativity. During a daydream, the brain is blending together concepts that are filed away in different areas of the brain, forming new connections that we call new insights. Without trying to focus, the brain starts to explore its inner database, looking for connections in a relaxed fashion.

Wide Attention—Relax

Yazulla

When trying to solve a problem, people often cover or close their eyes, reducing competing, non-essential input to the visual cortex. When the outside world becomes a distraction, the brain will attempt to block it out by selective attention. Messages with few or no distracting details are brain compatible with the nature of learning.

Whether to concentrate or relax depends on whether the information for the task-at-hand is at your fingertips or on the tip of your tongue. How quickly can you recall a piece of information when needed? Often, when a task is planned, detailed, and well-rehearsed (at your fingertips), concentration focuses your attention on the task. Such concentration may block out most distracting events around you. Sometimes, a piece of information is at the tip of your tongue and no matter how hard you try or concentrate, it just isn't there. It could be a name, a place, a spelling, song, just about anything that is usually familiar. You start going through the alphabet, a-b-c-d- . . . z, hoping to get a clue. Sometimes that works, sometimes it doesn't. It seems that the more you wrack your brain, the more distant the information becomes. So, finally you may give up, only having it pop into your mind once you stop thinking about it. In this case, concentration does not help, while diversion onto something else allows the mind to search your memory bank at leisure without the stress hormones released during intense concentration. Somehow stepping back from details, (i.e., wide attention) relaxes the mind for information to move more freely among neural circuits in the brain.

Hebron

I say to students, "Pay attention to what your attention is on, then receive or reject it." When learning, training, or performing, what is your attention on? ***It depends***. Our ability to pay attention is limited. When we pay attention to one thing, you ignore something else. Human attention is a resource and how it is used should be recognized.

There are ways in which attention is generated, manipulated, and valued. At times, attention is simply a lens through which we read the events of the moment. That process gives us some insights into how the emotional human mind works when learning, training, and performing.

When performing, some individuals are not accomplishing a wanted outcome because their attention is being used in ways that are different from when wanted outcomes are accomplished. For example:

○ Did the performer recognize their attention was on, "don't make a mistake," and not on what-to-do?

- ○ Did the performer catch their attention going in the direction of a past mistake and not the present act?

- ○ Did the performer recognize that the last performance cannot be fixed, and just moved on, or were they still thinking about it?

- ○ Did the performer recognize that their attention was leaving the present and drifting into the past or future?

- ○ Did the performer recognize that their attention was on judging every outcome and not just accepting them and moving on?

Diversify or Specialize

"It is better to know something about everything than everything about something."
—Adapted from Blaise Pascal

Yazulla

Contrast the above with the apparently opposing proverb "Jack of all trades, master of none." This is a portion of the original verse "A jack of all trades is a master of none, but oftentimes better than a master of one." The full verse of this saying dates back to the late 15th century and was used as a compliment for one who was a generalist as opposed to a specialist. Now, the shortened version has a negative connotation and is often used to describe a dabbler, a dilettante, someone with only superficial interest in many topics, but useful information in none. This is in opposition to the admired so-called "Renaissance Man," one with deep broad knowledge. As with all such proverbs, there is an element of truth in each view. Let us consider sports and then academia.

Hebron

Today, at every level of golf a prediction that a college coach made over 30 years ago has come true. Conrad Rehling, who coached golf teams at the University of Florida and the University of Alabama, said, "When the great athletes that are going out for football teams, basketball teams and baseball teams take up the game of golf, we will have more great golf." While advances in the technological areas of golf are not to be overlooked, in my view, it is the physical size and all-around athletic skill of the individuals who are playing golf today that has made the big difference in how the professional game is now played compared to 20 years ago.

Whether to specialize or diversify has been a contentious issue in sports. It is common today to see youngsters participating in multiple sports, sampling a variety as they are carpooled from one organized event to another. Youth teams in soccer, basketball, lacrosse, hockey, baseball, tennis, golf, self-defense, etc. allow youngsters to experience differences in the coordination of mental and physical demands of each sport. Depending on individual skill, motivation and interest, choices are made. As pointed out earlier, the choice the child makes for each sport is either: a) drop out, b) participate recreationally, or c) pursue more advanced skill. The question for parents of a child athlete has been either: year-round all-season training for a seasonal sport or to participate in several seasonal sports. Is the intense focus on a single sport helpful or not?

There has been much discussion and now a consensus on the positive value of multiple-sport versus single-sport focus in the development of overall athletic ability in youngsters. There are numerous examples of professional athletes who, in college and for pro teams played different sports; for example, Jim Brown, Bo Jackson, Dave Winfield, Deion Sanders, Bob Gibson, among many others, including Jackie Robinson and Jim Thorpe. Women, as well head this list including: an all-time great Babe Didrikson Zaharias, Lottie Dodd and Elyyse Perry. All celebrate excellence in multiple sports. Early specialization is associated with dropping out in sports, while staying involved supports early diversification.

A study of elite Russian swimmers found that fewer 9- and 10-year-olds who began specialized training were on their national team than the athlete who waited to begin specialized training until 13 or 14. These 9- and 10-year-olds also ended their sports careers earlier than athletes who started to specialize later in life (Bompa, 2000). A single focus on tennis at an early age contributed to withdrawal from the sport (Lochr, 1996). Parents of hockey players, both active players and ones that dropped out (ages 6–13), found the players who dropped out spent more time in deliberate, specialized practice in off-ice training (low enjoyment), than the expert athletes who experienced more free play (Hodges & Deakin, 1998). A lack of enjoyment was the most common reason for withdrawal from sports altogether (Ewing & Seefeldt, 1996).

Yazulla

Once a student enters college there is a conflict over how broadly the choice of required courses should be. A normal load in college is 120 credits (30 to 40 courses or so). A student will take 30 credits in their major, 30 credits in a related subject matter, and the remaining 60 credits are distributed in courses that should broaden the students' education. For example, a Biology Major will take at least 30 credits in Biology, 30 credits in a combination of Calculus, Chemistry and Physics. The other courses would be in so-called liberal arts, history, philosophy, literature, language, music, etc. The reverse is true for a History or Language major. Students of any major often complained about having to take courses completely out of the range of their major, particularly as the grades in these courses were weighted the same as major courses in computing their final grade-point average. This complaint falls on deaf ears because universities, professions and businesses recognize the value of people who know more about the world they live in than just a narrow piece of expertise. This attitude is under assault as colleges and universities are under pressure to prepare students for the job market; goal directed to marketable skills. The effect is to turn Universities into Trade Schools.

As a university Professor, I appreciated both views. There were students that had been goal-oriented to a profession since childhood. They were more likely to object to non-goal-oriented courses. Often, these students needed to be convinced to look at those courses as hurdles that had to be jumped before getting to their goal. My preference and advice to students was to cast a wide net. Introductory courses in most majors tend to be survey courses, covering broad areas of the subject matter. It is a good idea to sample majors through these introductory courses until you find something that captures your interest. Despite most students entering college with a major in mind, most students eventually change majors at least once before they graduate. I think this flexibility is healthy and prevents students from getting stuck in a course of study they regret. Diversification can aid in choosing a major, but once a major is chosen, the choice of Specialize or Diversify surfaces again. As a History major, do you want to specialize in ancient cultures: Medieval, Colonial, The Enlightenment, Eastern, African, Modern, and many others? As an Engineering major, do you like: Heavy engineering, Light engineering, Mechanical, Electrical, Chemical, Industrial, Civil, Biomedical, and so on? There are many choices even within a specialty.

> *"Overemphasis of the competitive system and premature specialization on the ground of immediate usefulness kill the spirit on which all cultural life depends, specialized knowledge included."*
> —Albert Einstein

My own experience may serve as a good example. My scientific specialization was to study the function of the retina (the part of the eye that processes light into nerve signals). The technique I started with was electrophysiology, the use of electronics to measure electrical activity in the nerve cells of the retina. Electrophysiology is a highly specialized technique, with a very large information base. However, in order to make sense of the data I was getting, I needed to know more about the structure (anatomy) of the retina and brain. I then proceeded to learn a variety of anatomical techniques: histology, microscopy, electron microscopy, autoradiography and then neurochemistry. As a result of broadening my experience, a new spectrum of problems became available to me that could be addressed with the new array of techniques I could bring to bear.

When I was a new Assistant Professor at Stony Brook University, a prominent neuroanatomist advised me against this strategy of diversification. His argument was that since I was so spread out by technique, I would not be considered an expert in any one technique and therefore, would not have a professional niche that I would be identified with. At that time in the late 1960s, the field of retinal science was broken down by technique: Anatomy, Electrophysiology, Biochemistry, Pharmacology, etc. In this regard he was right. I found myself at the periphery of these technique-oriented groups. I could participate in any of them, but I was not one of them. However, I found that the specialization strategy was incredibly boring and restrictive. Over the years, two groups developed that were technique oriented and those who were problem oriented. I was in the latter group and continued to choose projects of interest and use or learn the techniques that were needed to carry them out.

As neuroscience progressed, it became more and more interdisciplinary. Neuroscience drew in scientists from fields of study ranging from molecular genetics to artificial intelligence and everything in between. The result was that those who diversified their expertise were able to view problems from a broader perspective. Over the years, research papers, grant proposals and books that were multidisciplinary in content were sent for review and comment to those of us that had such diversified expertise. Scientists who remained specialists and did not diversify became techniques in search of a problem, but still they were sought out for their expertise until, for many, old techniques were superseded by the new.

Back to the question for a student, is it better to specialize or diversify. There is no doubt that any profession or job requires specialized expertise in its performance. In academia and science, it is often the extreme specialization or focus that leads to breakthroughs, as the scientist probes deeper and deeper into the problem. However, I think a generalist with an extensive information base who can step back with the information, see it, get an overview, and apply it in a wider context, is also critical to progress. Both have their place; both are essential. It seems to be up to one's personal preference which path to take.

Summary

"Focus" or "Multitask" When presented with multiple problems, it is more efficient to focus on one at a time than try to multitask, alternating by doing just a portion of each one. Depending on the difficulty, it takes extra time to shift attention and resources from one task to another. The brain simply does not do well attending to two competing inputs or complex actions at once.

"Narrow-Focus" or "Wide-Attention" Persistent, intense focus on any issue requires a lot of energy as demonstrated by fMRI imaging of the brain. Taking a break, letting your mind relax allows other memories, relevant and irrelevant, to intrude into consciousness. In such cases it is useful to step back from focusing on the bark on the tree, to realize that it is part of the tree, and the tree is part of a forest.

To Diversify or Specialize has much in common with whether to "Play" or "Drill." As with any start, it is important to sample many things to determine what you find of interest or fun, or even what to avoid. Once a general path is chosen, casting a wide net seems to be the best strategy because a broad training of body and mind allows a wide range of choices along the way, whether the activity is physical or intellectual. The choice of a generalist or specialist is very personal, for example, a family physician or cardiologist, decathlete or sprinter, general contractor, or electrician. In addition, there are specialties within specialties: general surgeon or vascular surgeon, dentist, or orthodontist. All are valuable; there is no single answer. It depends on one's interests and goals.

Your Notes

CHAPTER TEN

Reviewing Some Practical Solutions to Brain-Compatible Learning

"Don't hope that events will turn out the way you want; welcome events in whichever way they happen; this is the path to peace."
—Marcus Aurelius

Hebron

When it comes to coaching, in my opinion, there are many effective men and women helping and guiding their students. Each may have their way of supporting growth and development of skills, which can be a different way than other coaches. Every student is different, so is every coach.

An instructor may be a swing-the-arms coach, or a turn-the-body coach, or a technical coach or a non-technical coach, or a little bit of all those descriptions. Each coach may have their one go-to-suggestion that they feel will make the difference, which may be different from other coaches. Some coaches just accept a different way of coaching as that is your way and I have my way. Some coaches will criticize in a different way. Some coaches say they have the truth, when perhaps the truth is, *it depends*. And perhaps flexible thinking is the most effective coaching tool. Keep in mind what the great John Jacobs said, "there is some good in every golf book."

Hebron

I ask students if they want me to teach them or help them learn. I have never had a student say they wanted me to teach them. Keep in mind that when you teach, you can win or lose, but when you help someone to learn, you always win.

"Don't aim for perfection. Evolution and life only happen through mistakes"
—Matt Haig, English novelist

With what I call LEARNING POSITIVE GOLF™ (or your preferred topic), the first principle I pay attention to is emotion, then the student's past experience, and then subject content. We can gain insights that support learning from what is being uncovered about the brain's connection to learning.

- Learning is a personal, first-person experience

- Learning is emotional

- Learning is social

- Learning is biological in nature

- Learning is a chemical process

- Learning is internal

- Learning is mostly a sub-conscious event

- Learning is human development

- Learning is change

- Learning is supported by our past experiences

- Learning is supported by unworkable outcomes

- Learning is active, dynamic, and organic not passive

- Learning is a synthesis of connecting

- Learning is drawing out

- Learning is seeing options

- Learning is growth

- Learning is recognizing patterns and sequences

Is golf a sport or a game, an ego exhibition, or an experience? It depends on your view. Individuals, young and old, have been playing golf for hundreds of years. Over time many of these individuals were less-than-perfect golfers, but still found a way to enjoy their time on the golf course. On the other hand, some non-perfect golfers say they are not enjoying their time on the golf course.

Perhaps the non-perfect golfers who enjoy their game of golf have a different way of looking at golf than those who say they are frustrated and intimated when playing.

The non-perfect golfers who feel their time on the course is about how far they hit the ball and how low their score was, can let golf become an ego-driven exhibition. These individuals may see golf as a sport they have to beat. For them, golf is a test of their golf IQ, thereby emotionally influencing how they see their self-worth as golfers, or even worse, as a person.

On the other hand, the less-than-perfect golfers who see golf as a game to be experienced with fellow golfers have a cultural view that will support improving and enjoying their golf. It's participation over perfection that guides their view of the game of golf.

Unfortunately, there is a culture of golf that has always promoted perfection, which can lower enjoyment and future participation and performance. It's interesting to me that people can drive a few hours to go skiing, fall down several times and still say the experience was still enjoyable. On the other hand, after some golfers hit poor shots, they express all kinds of negative emotions.

It is emotionally unproductive to get frustrated with unintended outcomes when playing golf. Both experienced and less-than-perfect golfers may want to learn to accept what the game of golf is giving you on any day, including the workable and unintended outcomes. Playing golf within the spirit of the game, without trying to put on an ego-driven exhibition and enjoy a game that will be different every day, is the suggestion here.

Note: No golfer, professional or club golfer can predict their final score on the first tee, and every swing they make will have an outcome that may be not exactly as wanted, a reality that is often overlooked. Golf is inconsistent. Because of competition, getting a tour card or winning a tournament is hard, but playing golf is not. **The golf industry may want to stop saying golf is a hard game to learn and to play, and that may help grow the game.**

There are blind golfers, one-armed and one-legged golfers, and young golfers who can break 80. The game of golf should be recognized as inconsistent, not hard.

> *"Continuous improvement is better than delayed perfection"*
> —Mark Twain

Thoughts on Instruction

- ❍ Offer information that speaks to students, not at them.

- ❍ Be less formal than one might expect.

- Development has a history; it's built on memory.

- The fruits of decisions—ending up in a different place are a good thing.

- Inaccurate steps are always there—and beneficial.

- Use Possibilities—holistic in nature.

- Exercise restraint—while at the same time supporting the flow of allowing mentally and physically.

- Motion architecture is guided by the environment.

- Feel risky—push against formal structure.

- With freedom there is more influence over outcomes.

- The author of experience supports creative outcomes. (*The student*)

- Be creative, invert traditional rules.

- Self-trial and error learning—uncovers what you want.

- Help Students Invent.

Yazulla

What would I have done differently? Unlike Michael, my professional teaching career is over. Still there are opportunities to pass on information as with my grandchildren, or others in a social situation, or even in this book. As I look back on my teaching career, (I retired in 2013) and my collaboration with Michael over the last 9 years or so, would I have done anything differently? Yes, but that would have required me to change some of my expectations.

Teaching Biology Majors has a particular intensity in that a very high percentage of students have a goal of attending Medical School. The motivation is high grades and the information needed to do extremely well on the MCATs, the standardized test required for Medical School application. A textbook for any course in Biology: biochemistry, cell biology, genetics, physiology, neuroscience, etc. contains numerous chapters and up to 1000 pages. This is far too much material for a single course or often two courses to be comprehensive. Often, general topics of cellular and organ systems physiology are covered because these form the foundation for more advanced course. Still, the result is that a great deal of information is covered in a limited amount of time with the burden for details on the student. There is little time for in-depth discussion as students often find these diversions detract from "Just what do I have to know." And of course, I fell victim to this attitude in the beginning and loosened up over the years as I got more experience.

In retrospect, I would have spent more time on a general principle of all living organisms and less time on the numerous organ systems that exemplify that principle. All living organisms have to solve many examples of one basic problem: how to get a "something" across a barrier. That something can be water, air, or fat and whatever is contained therein: salts, oxygen, vitamins, etc. The barrier is usually the membrane that surrounds the cell. The membrane separates the interior of the cell from the environment. Once these principles are clearly understood, the details of the different organ systems: heart, lung, kidney, neuron, etc., are more easily placed into context.

As pointed out, I usually had all the answers and in the interest of time, I usually provided them. If I did not know the answer, I would find out and report at the next class. The rate of information flow often precluded extended questions and answers. I could sense the students wanting to move on from a sticking point—just get on with it. There were so many intriguing issues in physiology that were addressed in later bull sessions rather than the classroom. It was the race to cover as much material as possible within the limits of the lecture time that I found most hectic and disappointing. It would be good to revisit those days in which there would be more of a relaxed, student-faculty give-and-take. You never know what you might learn.

My favorite lesson to me was during a class discussion of hormones on embryonic development. A student, who happened to be a dog breeder, told the class about experiments in the Soviet Union in the early 20th century, regarding the domestication of foxes. By selecting the most docile kits in a litter and then breeding them, a domesticated fox was the result in five-six generations. At a later time in a Departmental seminar, this tidbit of "fox domestication" was relevant to a seminar topic of early brain development. I posed my comment and question to the speaker. A colleague leaned over to me and whispered, "Why do you know that?" I turned my head and said, "I learned it from a student in class." What would I do differently? Slow the pace, present less, ask more, and listen more.

"Experience is what you get, when you don't get what you want."
—Randy Pausch

Hebron

Treat unintended outcomes as desirable developmental difficulties. After I heard Dr Bjork say, desirable difficulties, I add the word developmental to his statement.

Learning is a social and emotional developmental process. Brain-Compatible Approaches to learning have demonstrated that the most effective way to evaluate students or for students to evaluate themselves is by moving away from seeing unintended outcomes as failure, in need of fixing. They are over, they can't be fixed. Unintended outcomes are valuable feedback for future use. They are more useful for experiencing deep learning than wanted outcomes. The feeling of a mistake can help you learn something different. An unintended outcome provides an opportunity to adjust and adapt.

No one is perfect; we are all human. Unintended outcomes are intrinsic to the human condition. They are a biological necessity for meaningful learning. A Huge insight. Failure 101 is an actual freshman course at Penn State University, under Professor Jack V. Matson, where the value of failure is explored. Unfortunately, the business of instruction often backs away from the value of poor outcomes. Flaws should and could be seen as undeveloped outcomes on their way to success.

When a student swings and the outcome is workable, I suggest to them that they have shown they are capable of that workable outcome. Now, moving forward I suggest looking at training as gaining some understanding of what gets in the way of accomplishing that outcome. The obstacles may be their mental concepts and emotions, and not their ability to create wanted outcomes.

This is the reasoning I use. Telling students that they are doing something wrong is not as useful as making them aware that what they are doing is a journey of growth and development for what to do differently. The latter attitude is a positive approach, the former is not.

Do not see an unintended outcome as something that needs to be fixed; it is over; it is done. It cannot be changed. Pay attention to what-to-do not what-to-fix. If you burn a pie while baking, it can't be fixed, but you can bake the next one differently. The same holds true for completed golf shots; they cannot be fixed. After a hooked shot, a positive suggestion may be, "Do you think having the club face more open at impact for the next swing would change ball flight?" rather than "we have to fix that closed club face." The same information just presented differently.

The following are some of the things that I have said to students during instruction for improving long-term learning.

- ○ Look for workable not perfect on your journey of growth

- ○ Realize that a poor outcome cannot be fixed, it is OVER.

- ○ Pay attention to what to do, not what went wrong

- ○ Understand that ball flights cannot be controlled; if they could, pros would not miss shots. You can control what goes on in your mind before you address the ball.

- ○ Poor shots are part of the game and being human.

- ○ See me as a partner on a journey of growth. I do not have the answer; you do.

- ○ See unintended outcomes as valuable for learning what to do differently.

Block Practice

Doing something over and over may support progress during training, but random training supports long term learning and the retrieval of skills beyond short term performance. When training your golf swing, create random outcomes, high, low, left, right ball flights. Mixing new information with prior knowledge becomes new learning and a memory that can be recalled. For example, first making a hard-fast swing, then a slow-golf swing supports learning the feel of the speed in the middle or smooth golf swing. The brain then uses the differential *feel* of swings as reference points for learning.

Golfers say they want a consistent swing, but think again, we never have the same shot when playing. Suggestion: develop flexible, portable swings that can produce different ball flights. Avoid the popular block practice approach that does not prepare you for the ever-changing conditions the game of golf presents.

Learning Aids

Studies by Hodges and Campagnaro (2012) noted that when helping someone learn a golf or tennis swing, someone or something may be trying to physically guide the learner through the desired motions. Physically guiding may reduce errors when training but when playing, that guidance can no longer be relied upon, and meaningful learning suffers. Learning aids cannot be used when playing and therefore they are out of context. Studies at UCLA's Learning Lab show the biggest problem with aids is that they remove struggling, our most useful tool for learning.

Self-defined movements are more resistant to forgetting than when movements are physically controlled by others and aids. This is one of the most robust and reliable discoveries in the motor-learning literature.

Yazulla

I found that three important learning aids in academia are: a) a pencil, b) another student, and c) a good night sleep. As a learning aid, nothing beats taking pencil and paper in hand, and physically writing down, in outline form, the information you are trying to learn. As mentioned, simply reading a textbook as a novel does not work. Typing is not the same as writing. Writing slows you down and involves multiple sensory-motor pathways to reinforce new information. Another student, testing yourself by writing from memory is good, but nothing reinforces retention of new information like explaining it to someone else, then, have them explain it back to you. You can rehearse in your mind, but often you skip over details thinking, "Yeah, I know that" and you move on. Before computers, notes were handwritten, and we always formed study groups. Now, many students if they take notes at all, use a computer and study alone. The reason?—the fear that the other student will benefit and get a better grade. A **pencil** and **another student**—so many academic problems would be solved. Finally, get a good night sleep. As mentioned in more detail earlier, all-nighters do not work. Previously learned material is rehearsed during deep slow wave sleep. The physical and mental toll of poor sleep or no sleep will interfere with any possible gain of an increased time studying. An all-nighter may work if you start out knowing absolutely nothing, but it is a poor strategy for success.

Context

Hebron

When training anything, changing the context of the environment every 5 or 10 minutes supports long-term learning more efficiently than training in one environment for the entire time you are training (Goode, et al., 1986). Hundreds of experiments have demonstrated the value of this spacing effect as highly robust and reliable for learning to solve new problems (Cepeda, 2006).

Feedback

Hebron

A common assumption is that frequent feedback from an external source during the acquisition stage fosters long-term learning. But empirical evidence suggests that delaying, reducing, and summarizing feedback is better for long-term learning (Schmidt & Wulf, 1997). Frequent feedback can be distracting in that it can break the student's concentration. The student is attending to what the instructor is saying or will say rather than what he/she is trying to do. The feedback becomes a crutch during training. But that feedback is no longer present during a performance. Those frequent "atta boys" become associated with the learning paradigm, and their absence is felt during a performance. Golfers take note, having feedback after each swing can hurt growth and self-development.

Yazulla

Feedback in academia takes the form of graded quizzes, tests, or exams, in increasing order of time in class and content covered. The more time spent in class taking tests, the less time there is for instruction. Too many tests take away from lecture time, putting the burden on the student for the required course content. Too few tests' places much too much weight on any one test for the final grade. I opted for four tests: two quarterlies, a midterm and comprehensive final. In my experience, by the second test, most students had shown their level of competence with the course material. Of course, there were some students who showed marked improvement over the semester. It was really for those students that I maintained a 4-test regimen. Also, a comprehensive final exam gave students a chance to improve over a weaker performance in an earlier test. I felt it was important to give a failing student a chance to pass the course by doing well on the final. Why even try if there was no hope?

Time Distribution

Hebron

If you have just one hour to train, it is not as useful to train for that one hour straight as it would be to break the hour into four 12-minute sessions, with breaks in between. Also training something different in each 12-minute session is more useful than training the same thing in each session. Some golfers spend 90% of their time on their swing. The swing is less than 10% of what Jack Nicklaus, Tiger Woods and other great players have said are the elements that support learning and playing golf up to their potential. These 20-plus elements factor into a peak performance: Imagination, emotional control, mental focus, physical conditioning, adjustable skills, reasoning powers, reading greens, picking clubs, organizational skills, memory, self-discipline, past experience, self-confidence, training schedule, creativity, determination, self-control, anticipation, mindset, work ethic, and so on.

Keep in mind that golf swings are not up to their potential all the time, and great golfers play golf not golf swing. In fact, many tournament winners have said that they DID NOT have their swing on some weeks they won. Yes, the golf swing is important and one of the many elements that support learning and playing up to one's potential.

Swings without staying in the present or having your thinking focused may not bring the best outcome. Many swings that are admired on tour and at local clubs often do not outscore less-admired swings.

Yazulla

"Chunking," there are two ways of looking at chunking. First, chunking is a strategy that breaks up a complex project into discrete parts (chunks) that each are manageable and less intimidating than tackling the entire project as a whole. Chunking is an efficient strategy for so many tasks we encounter, like packing to move, clearing out a garage or attic, painting your house, furnishing a house, planning a wedding, and so on. Breaking a task down into manageable portions gives you a sense of accomplishment when any of them is completed. Second, chunking can be looked at in the reverse that is combining related items into a single concept. Chunking is a valuable strategy in academia, in which steady progress provides a sense of relief that is realized at exam time.

A problem I found with so many students was that they thought they could study chapters in a physiology textbook the way they would read a novel. Just read through the chapter, and that was adequate enough. Except for those of us without photographic memory, it does not work that way. Chapters are composed of episodic tidbits of information, though related around a topic, they are not continuous as you would find in a story. For example, a chapter on the Heart would have a section on heart anatomy, then structure of heart muscle, blood supply of the heart, nervous control of the heart and so on. Though these topics are all related to heart function, they differ in vocabulary and concepts. My advice was to chunk-out individual sections and concentrate effort on each section at a time.

Every time you open that textbook, have a purpose in mind such as "This time, I am going to learn about cardiac muscle." Open the book, deal with this topic alone and stop. Now review it. Take a break. Your brain needs some time to consolidate the new information. Do not jump to the next topic. This tactic can lead to a disruption of the transition from short-term to longer-term memory of what you have just read. I speak of this from personal experience of thinking I could pass an exam just by reading through the textbook. I was lucky enough to get copies of previous final exams. To my horror, I was clueless about details in that exam. That is when I learned to focus on a specific topic and concept, learn it, rehearse it, and then move on to the next topic.

Alternatively, chunking involves combining multiple related parts into a coherent concept. For example, using the heart again, the explanation of why blood pressure changes as blood moves through the circulatory system makes sense using an analogy of water pressure in a garden sprinkler and soaker system. The individual elements are then viewed as parts of a single concept.

One of the most difficult things for students to learn is time management. For many, it is time away from home, freedom, no one checking up on them. Students get distracted by the extracurricular opportunities and lose track of the time frame to take care of their academic obligations. Eventually, it is exam time, and for some that is too late. Effective time management and chunking is a very effective way to avoid cramming that is the price to be paid for procrastination

Starting a Session

Hebron

I normally let students, pros and amateurs know that I believe that they are not broken, in need of fixing. They are just on a journey of development to discover what can be done differently, not better. Better does not exist in the brain. If I miss a putt and make the next one, it was not better; it was different, more to the right, to the left, shorter, etc. The brain operates by making predictions, if this, then that, based on prior experiences, workable and unworkable.

If working on 10-foot putts, hit some long, then short and the brain will calculate the middle, or 10-foot sub-consciously. To find 10 feet, the brain needs to know what longer and shorter of that distance is, so it can make a distance prediction. Also, have students swing too fast, then swing slow, now find the middle. This comparative approach is how the brain learns, not by trying to get anything right.

The time spent with me is a session not the lesson, what the student takes away from the appointment is the lesson. I tell students that our time together is a time to experiment, not to get something right. I tell students that both workable and unworkable outcomes are valuable for learning. I tell them that I am their coach and that the outcome of their swing, workable and unworkable is their self-teacher.

The late Fred Rogers (1928–2003) summed it up nicely.

> "If we grow up fearing mistakes, we may become afraid to try new things. Making mistakes is a natural part of being human and a natural part of the way we learn. It's an important lesson, at any time of life, but certainly the earlier the better. We all make mistakes as we grow, and not only is there nothing wrong with that, but there's also everything right about it. (Rogers, 2006, p.156)

Your Notes

Appendix

Basic Information on the Nervous System

The following section introduces the major players of the brain and nervous system, along with environmental conditions in which the normal, healthy brain functions.

The Nervous System

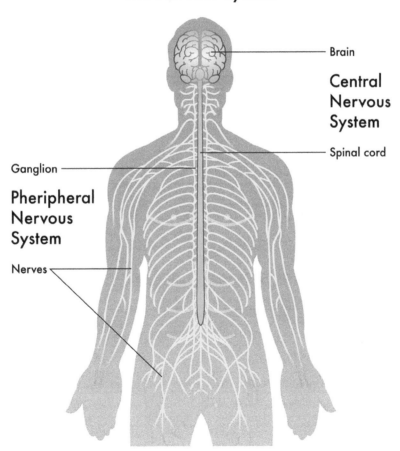

FIGURE 7: *The schematic illustrates that the brain and the spinal cord comprise the central nervous system (CNS). All other nerves of the body, including those of the sense organs, muscles, glands, and internal organs comprise the peripheral nervous system (PNS).*

The human brain weighs about 1 pound at birth and reaches its final weight of about three pounds by 10–12 years of age. During the rapid growth phase in infancy, the brain adds about 250,000 nerve cells per minute by cell division (mitosis).

The brain contains two major types of cells: neurons and glia.

Glia. (Latin, meaning glue) are supportive cells that surround neurons and are responsible for maintaining the chemical environment of neurons. Glia control the balance of water, salts, sugars, and other chemicals that modulate the activity of neurons.

When one thinks about the brain, one often thinks of the 'old gray matter,' i.e., neurons. Amazingly, neurons account for only 10% of the cells in the brain; 90% of cells in the brain are glia. Even though neurons are outnumbered 10:1 by glia, neurons occupy 40% of the brain volume.

Along with an extensive network of blood vessels, glia provide the critical function of house-keeping: energy supply, waste removal, growth, and repair; they protect neurons from substances in the blood (Blood-Brain Barrier), respond to infections, inflammation, and so on.

Structure of a Typical Neuron

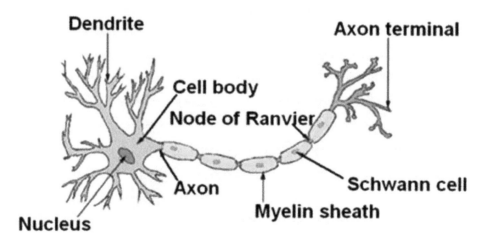

FIGURE 8: *Neurons are specialized cells that transmit information by chemical and electrical signals. Input to the neuron ordinarily occurs at branched structures called dendrites (Greek for branch). The output is transmitted along a wire-like structure called an "axon" by an electrical signal. Neurons communicate with each other by releasing chemicals the axon terminal at structures called "synapses" (Greek, for 'joining together'). The chemical travels a very small distance (less than one-hundred thousandth of an inch) to stimulate the dendrite of the next neuron in the sequence.*

Chemical Messengers

Any single neuron can make synapses with only a few, or many thousands of neurons. Chemicals released at synapses are called neurotransmitters, and they act locally over very short distances, less than one-thousandth of an inch. These transmitters bind to specific receptors on the receiving neuron to produce a response that then is transmitted to the next neuron in the network. Neurohormones are chemical messengers that can act over longer distances or on organs throughout the body. Neurohormones may be released by neurons locally, or specialized cells in endocrine glands that release the hormone directly into the bloodstream. Two examples are adrenaline and steroid hormones.

Generic Neurotransmitter System

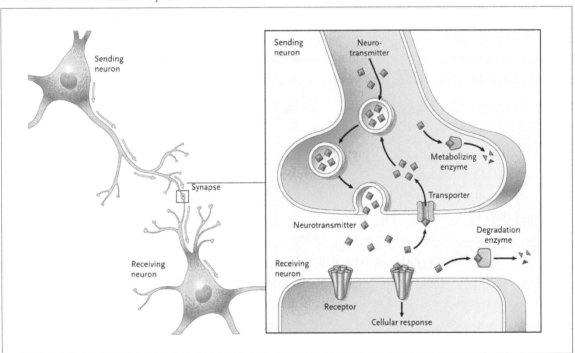

FIGURE 9: *Illustrates two neurons (left), one sending, the other receiving information at a specialized structure, a synapse. The enlargement of the synapse (right) shows neurotransmitters stored in vesicles and released into the space between the neurons. The neurotransmitters bind to protein receptors and activate the receiving neuron. The neurotransmitters are then degraded and recycled. The sequence repeats itself at the next stage in the nerve network.*

Some neurotransmitters make neurons more active (excitation), while others make neurons less active (inhibition). Many neurotransmitters can be excitatory or inhibitory, depending on the target tissue. For example, acetylcholine stimulates skeletal muscle and also inhibits heart muscle but by acting on different types of receptor proteins. The whole nervous system operates by a delicate balance between excitation and inhibition. The nervous system is never at zero activity. It is always in a steady state that can be excited to do more or inhibited to do less.

For example, motor neurons in the spinal cord that stimulate skeletal muscles (like in the arm or leg) receive a steady stream of excitatory input from the brain. Thus, at rest, skeletal muscles are in a constant state of tone. Even when relaxed, they are under some tension. In contrast, think of the dead weight your leg feels like if it falls asleep. Cutting off the blood supply to the leg or arm stops the chemical stimulation of the muscles, and they go limp. Similarly, the venom of a cobra blocks the excitatory chemical stimulation of the skeletal muscles, particularly those of the diaphragm that control breathing. The result is paralysis and suffocation. Whereas a rat poison (strychnine) blocks the inhibitory input from the brain to spinal cord motor neurons. The result is that there is excessive and unregulated excitation of skeletal muscles, leading to spastic convulsions and death. In each case, the balance of excitation and inhibition to the skeletal muscles was upset, resulting in either flaccid paralysis (cobra venom) or spastic paralysis strychnine, (rat poison).

Neurohormones that enter the bloodstream travel throughout the body acting over large distances, affecting neurons and cells throughout the brain and organs in the body. For example, if you are startled or frightened, you freeze, may feel a flush, a rush of heightened attention and alertness. This 'fight-or-flight' reaction is caused by a release of adrenaline and nor-adrenaline from cells in the brain and adrenal gland. The effect of these hormones is to increase lung and heart activity, blood flow to skeletal muscles, mental alertness, and attention to whatever caused the startle response. At the same time, there is suppression of the entire digestive system. This is one reason that if you engage in strenuous activity after eating it can cause indigestion. There is not enough blood in your body for all of your organs. So, blood that would be needed for digestion is diverted to your skeletal muscles and the result is that the newly eaten food just sits in your stomach, causing gastric distress. Adrenaline and other stress hormones have enormous effects on the brain and cardio-vascular activity. These stress hormones affect attention, alertness, memory, and sense of well-being, all of which are important factors in learning situations.

Protection of the Brain Space

The adult human brain has about 100 billion neurons and one trillion (1,000,000,000,000) glia. Mixed among all these cells is an extensive web of blood vessels that supply nutrients (oxygen, glucose, amino acids, minerals, vitamins) and carry away waste (carbon dioxide, ammonia). In a normal condition, blood cells and neurons never mix. Blood cells are deadly to neurons. Head trauma and strokes in which blood leaks into the brain can cause rapid death of neurons. Blood and neurons are kept separate by an elaborate system called the "Blood-Brain Barrier."

For example, 24 oz of a cola contain about 78 grams of sugar 3 ounces, 18 teaspoons. Shortly after drinking the cola, your blood sugar will show a rapid increase, but the sugar content of your brain environment will change very little. The blood-brain barrier controls the concentration of sugar, salt and other substances that dissolve in water within very narrow limits to protect the brain from changes that occur during the day. However, substances that dissolve in oil (fat soluble), for example, nicotine, cocaine, or general anesthetics, pass right through the blood-brain barrier with potentially great effects on brain activity.

Brain Regions Have Different Functions

The 100 billion neurons are not all the same; they are of different types, and they are not uniformly distributed throughout the brain. The idea that specific areas of the brain are responsible for specific functions is well understood today, but only has been accepted for the last 150 years. (Franz Joseph Gall 1758–1828) made the first attempt to study brain function which is described in his paper *The Anatomy and Physiology of the Nervous System in General, and of the Brain in Particular.* Gall's idea was that enhanced faculties of the brain would be expressed as expansions of the brain and would be visible as protrusions or bumps on the surface of the skull. This gave rise to the popular rage of Phrenology. Ceramic busts with these brain faculties indicated were popular and could be purchased at any Pharmacy. Although long discredited, Gall's efforts drew attention to the ideas that different areas within the brain were responsible for specific functions.

With advances in medicine in the 19th century, more people survived brain injuries, whether due to disease, accidents, or war. These provided opportunities to study the effect of brain injuries on behavior and thereby infer the area of the brain involved in particular functions. For example, in the 1860s language areas of the brain on the left hemisphere were identified by autopsy of patients who had severe problems expressing language (Broca's aphasia) or understanding language (Wernicke's aphasia). The situation regarding language is more complicated than indicated by these early studies. Still, the stage was set for the study of the functional anatomy of the brain, that is, brain and behavior.

Sir David Ferrier (1843–1928) published *The Functions of the Brain* in 1876. Over time studies into the brain would become less general, with specific topics including the brain's connection to learning being researched. The brain has two hemispheres, right and left, that appear identical to any casual observer. However, functionally, they are different. In general, the left hemisphere controls sensory inputs and motor outputs for the right side of the body. The reverse is true for the right hemisphere. It is common knowledge that someone who has had a stroke in the left hemisphere is affected on the right side of the body, usually with some degree of paralysis. The two hemispheres are connected by a large band of nerve fibers called the Corpus Callosum (from Latin, 'tough body') that allows the two hemispheres to communicate with each other. In this way, the brain functions seamlessly as an integrated unit instead of separate structures.

The outer, visible part of the brain is the highly convoluted cerebral cortex. It is by use of the cerebral cortex that we think, reflect, and communicate with language and abstract symbols—functions that make us human. Other mammals have a cerebral cortex, but

there are qualitative differences among us that separates humans from them. What this difference is we can only speculate; but it is not important for our purpose here.

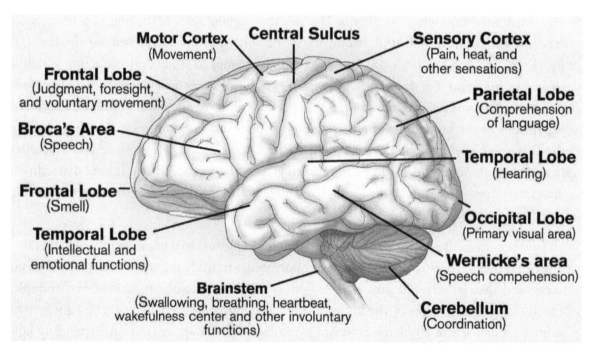

FIGURE 10: *There are functional differences among brain areas in the cortex. The visual sense is coded largely in the back of the brain; hearing on the sides, just above the ears; language ability, both understanding and expression, is on the left side; skin senses and skeletal muscle control—halfway between the front and back of brain; planning, intention, initiation of action, control of impulsive behavior (so-called executive functions)—in the front of the brain.*

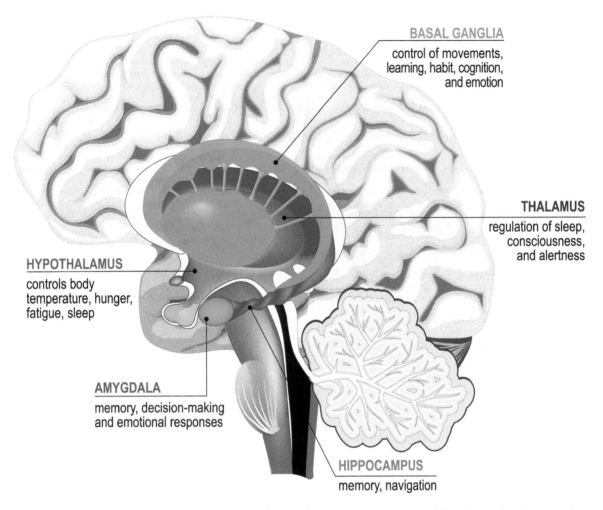

BASAL GANGLIA
control of movements,
learning, habit, cognition,
and emotion

THALAMUS
regulation of sleep,
consciousness,
and alertness

HYPOTHALAMUS
controls body
temperature, hunger,
fatigue, sleep

AMYGDALA
memory, decision-making
and emotional responses

HIPPOCAMPUS
memory, navigation

FIGURE 11: *As illustrated, there are several specialized structures covered by the cerebral cortex that are important for our discussion of learning, performance, and emotion:*

○ Hippocampus—is critical for the formation of memories.

○ Amygdala—is critical to assess the threat level or lack thereof of any incoming stimulus. The filter for the initial 'flush' before a threat is identified.

○ Basal ganglia—for coordination of skeletal motor control. Parkinson's disease is a result of damage to the basal ganglia.

○ Hypothalamus—thirst, hunger, body temperature; controls hormones involved in metabolic systems, growth, reproduction, immune system, kidney function and so on. Controls the pituitary gland and is really the "Master gland."

○ Limbic system—is an ancient part of the brain that is involved in primitive attributes of emotion, rage, and sex.

○ Hindbrain—connects the spinal cord with the rest of the brain; it includes:

○ Cerebellum—critical for fine motor control and so-called "motor memory"

○ Medulla—controls breathing rate, heart rate and vegetative functions.

As you might imagine, specific behaviors are correlated with increased activity in different brain regions as determined by functional Magnetic Resonance Imaging (fMRI). When you speak, the language areas in the left side of the cerebral cortex are more active; but when you are reading aloud, the visual areas in the back of the brain become more active. If you are walking at the same time, then the motor cortex on both sides of the brain gets involved. The Frontal Cortex has been directing this activity all the time, as well as the cerebellum that coordinates the automatic behaviors of balance and so on. As you are walking, your environment is evaluated for potential threats by the Amygdala and responses to these by the Limbic System and Hypothalamus.

The Changing Brain

As mentioned, the brain triples in weight from birth to adulthood, adding tens of billions of neurons during this time. All of these new neurons must be in the right region of the brain and connect with the appropriate targets. Chemical signals guide the rate of growth and direction of growth for the dendrites and axons of these neurons. Two examples below illustrate this important point.

Axons of sensory neurons in the hand must find their way to synapse on the proper neurons in the spinal cord, and these neurons from the spinal cord then grow into the brain. For movement to occur, the neurons in the brain must find their way into the spinal cord, whose motor neurons must go to the proper muscle to be activated.

Neurons in the eye must respond to the proper stimulus in space, find their way into the proper part of the brain to form an organized map of the visual world. These neurons in the brain must respond to the appropriate shape, size, color, and position of a stimulus, such that the person can say not only "I see it" but "what is it," "where is it," "new or familiar," and "friend or foe."

Even though the growing brain adds neurons, not all of the neurons present in a child's brain survive into adulthood. Why? Simply stated, the new and developing neurons are like small seedlings. As they grow, they sprout dendritic branches, increasing greatly in number, complexity and volume.

These branches compete with each other for the limited space in the growing brain. Neurons that are activated together will form strong synaptic connections and survive. Neurons that are not activated by stimuli that are close in time or space do not strengthen synaptic connections, and these neurons eventually die. This gives rise to the well-known phrase in developmental neuroscience "Neurons that fire together, wire together." Inactivity also can weaken existing synaptic connections—"Use it or lose it."

Sensitivity of the developing brain

For the sensory systems of touch and vision, there is a critical time in development for stimulation to form the proper neuronal connections in the brain. For example, babies born with cataracts (clouding of the lens in the eye) will be permanently visually impaired if removal of the cataract is delayed until late childhood. Also, notice how easily children learn language and how difficult it is to learn an additional language as an adult. This is because the developing brain is more dynamic plastic at a young age and becomes more hard-wired as we get older.

The developing brain is very sensitive to its chemical environment. During pregnancy, consumption of alcohol, cocaine, amphetamine, nicotine, etc., by the mother can have serious effects on brain development of the fetus. This is because all these drugs get past the blood-brain barrier and have the same effects on the fetus as they do on the mother. Each of these drugs affects specific chemical transmitter systems and metabolism in the brain. The proper timing and coordinated operation of these transmitters is critical for proper brain development.

Development and Learning

Learning is similar to development in that learning also involves and requires changes in the synaptic organization of the brain. The changes that occur with learning are not as rapid or extensive as that occurring during development. However, the ability of synaptic changes to occur in both conditions is influenced by the chemical composition of fluids bathing the neurons. The chemical medium of the brain is influenced by what we eat, drink, breathe and put on our skin, as well as by our physical and social environments. The public and medical communities are aware of the great attention paid to factors affecting the brain as they relate to prenatal care and fetal development. However, less attention has been paid to these factors as they relate to any Learning Environment.

A learning situation that takes effects on the brain into account so as to maximize learning is referred to as "Brain-Compatible." The brain is not a black box that is to be filled or tapped at will without considering the effects that the teaching methods or learning environments have on the brain. Stress, fatigue, diet, past experience, and expectations are among the numerous factors that can affect the chemical environment of the brain during learning. Understanding the source of these factors to minimize negative effects on learning promotes a Brain-Compatible Approach.

Bibliography

Able, T., Haveks, R., Saletin, J.M. & Walker, M.P. (2013). Sleep, plasticity, and memory from molecules to whole-brain networks. *Current Biology, 23*, R774-R788.

Alred, D. (2016). *The Pressure Principle: Handle Stress, Harness Energy, and Perform When It Counts.* Penguin, ASIN: B017N4YVZA

Aslaksen, K., & Loras, H. (2018). The modality-specific learning style hypothesis: a mini review. *Frontiers in Psychology, 9*, 1–5. Doi: 10/3389/fpsyg.2018.01538.

Beets, I.A.M., Macé, M., Meesen, R.L.J., Cuypers, K., Levin, O. & Swinnen, S.P. (2012). Active versus passive training of a complex bimanual task: Is prescriptive proprioceptive information sufficient for Inducing motor learning? *PLoS ONE, 7*(5) e37687. https://doi.org/10.1371/journal.pone.0037687

Berry, J., Abernethy, B. & Côté, J. (2008). The contribution of structured activity and deliberate play to the development of expert perceptual and decision-making skill. *Journal of Exercise and Sport Psychology, 30*, 685–708.

Bompa, T. O. (2000). *Total training for young champions.* Human Kinetics.

Bouchard, A.E., Corriveau, H. & Milot, M-H. (2015). Comparison of haptic guidance and error amplification robotic trainings for the learning of a timing-based motor task by healthy seniors. *Frontier Systems Neuroscience, 9*(52), https://doi: 10.3389/fnsys.2015.00052

Bransford, J. (2007) *Understanding the Brain: The Birth of a Learning Science*, OECD

Buckingham, M. & Goodall, A. (2015). Reinventing performance management. *Harvard Business Review*, April.

Cepeda, N. J., Pashler, H., Vul, E., Wixted, J. T., & Rohrer, D. (2006). Distributed practice in verbal recall tasks: A review and quantitative synthesis. *Psychological Bulletin, 132*(3), 354.

Chiyohara, S., Furukawa, J., Noda, T., Morimoto, J., & Imamizu, H. (2020). Passive training with upper extremity exoskeleton robot affects proprioceptive acuity and performance of motor learning. *Scientific Reports, 10,* 11820. https://doi.org/10.1038/s41598-020-68711-x

Corbett, M. (2015). From law to folklore: work stress and the Yerkes-Dodson Law. *Journal Managerial Psychology, 30*(6), 741–752. https://doi:10.1108/jmp-03-2013-0085

Darden, G.F. (1997). Demonstrating motor skills—Rethinking that expert demonstration. *Journal of Physical Education, Recreation and Dance, 68*(6), 31–35. https://doi: 10.1080/07303084.1997.10604962

Davachi, D. & Dobbins, I. (2008). Declarative memory. *Current Directions in Psychological Science, 17*(2), 112–118. DOI:10.1111/j.1467-8721.2008.00559.x

Diamond, D. M., Campbell, A. M., Park, C. R., Halonen, J., & Zoladz, P. R. (2007). The temporal dynamics model of emotional memory processing: a synthesis on the neurobiological basis of stress-induced amnesia, flashbulb and traumatic memories, and the Yerkes-Dodson law. *Neural Plasticity,* 60803. https://doi.org/10.1155/2007/60803

Dowrick, P. W. (1976). *Self modelling: A videotape training technique for disturbed and disabled children.* Doctoral dissertation, University of Auckland, New Zealand.

Dowrick, P. W. (1999). A review of self-modeling and related interventions. *Applied and Preventive Psychology, 8*(1), 23–39.

Ewing, M.E., & Seefeldt, O. (1996). Participation and attrition patterns in American agency-sponsored youth sports. In F.L. Smoll, & R.E. Smith (Eds.), *Children and Youth in Sports: A Biopsychosocial Perspective,* (pp. 31–46). Brown and Benchmark.

Furey, W. (2020). The Stubborn Myth of "Learning Styles"—State teacher-license prep materials peddle a debunked theory. *Education Next, 20*(3), 8–*12.*

Giradeau, G., Benchenane, K., Wiener, S.I., Buzsáki, G. & Zugaro, M.B. (2009). Selective suppression of hippocampal ripples impairs spatial memory. *Nature Neuroscience, 12,* 1222–1223.

Goode, S. & Richard A. Magill, R.A. (1986). Contextual interference effects in learning three badminton serves, *Research Quarterly for Exercise and Sport, 57*(4), 308–314, https://DOI: 10.1080/02701367.1986.10608091

Hebron, M. (2001). *Golf Swing Secrets . . . and Lies.* Learning Golf, Inc.

Hebron, M., Yazulla S. Collaborator, (2017). *Learning with the Brain in Mind: Mindsets before Skillsets.* Learning Golf, Inc. ISBN 978-1-937069-08-7

Hobbs, B. & Artemiadis, PP. (2020). A review of robot-assisted lower-limb stroke therapy: unexplored paths and future directions in gait rehabilitation. *Frontiers Neurorobiotics, 14*(9), https://doi:10.3389/fnbot.2020.00019

Hodge, T., & Deakin, J. M. (1998). Deliberate practice and expertise in martial arts: The role of context in motor recall. *Journal of Sport and Exercise Psychology, 20*, 260–279.

Hodges, N.J. & Campagnaro, P. (2012). Physical guidance research: assisting principles and supporting evidence. In N.J. Hodges and A.M Williams eds.), *Skill acquisition in sport, Research, theory, and practice.* (pp. 150–169). Routledge.

Holyoak, K.J. & Morrison, R.J. (eds). (2013). *The Oxford Handbook of Thinking and Reasoning.* Oxford University Press. ISBN-13: 978-0199313792

Huber, R., Ghildari, M/F., Massimini, M., & Tononi, G. (2004). Local sleep and learning. *Science, 430*, 78–81.

Immordino-Yang, M.H. (2016). Why emotions are integral to learning. Excerpted from Immordino-Yang, M.H. November (2015) *Emotions, learning and the brain: Exploring the educational implications of affective neuroscience.* W.W. Norton & Co.

Jones, R., Armour, K.M., & Potrac, P. (2014). *Sports Coaching Cultures; From Practice to Theory.* Routledge. ISBN-13: 978-0415328524

Kensinger, E. A., Clarke, R. J., & Corkin, S. (2003). What neural processes support encoding and retrieval? An fMRI study using a divided attention paradigm. *Journal of Neuroscience, 33*, 2407–2415.

Knowles, M. (1983). *Self-Directed Learning, a Guide for Learners and Teachers.* ISBN-10: 0842822151

Lembke, A. (2021). *Dopamine nation: finding balance in the age of indulgence.* Dutton. ISBN-10: 0670785938

Leher, J. (2012). *Imagine: How creativity works.* Houghton Mifflin. ASIN: B007QRI1UQ

Levine, M. (2002). *A Mind at A Time.* Simon and Schuster. ASIN: B0012FBA38

Lochr, (1996). Cited in Hebron Podcast Golf Science Lab,

MacDonald, M. (2008) *Your Brain: The Missing Manual,* O'Reilly Media. ISBN-10:0596517785

McCullagh, P. and Caird, J. (1990). A comparison of exemplary and learning sequence models and the use of model knowledge of results to increase learning and performance. *Journal of Human Movement Studies, 18*, 107–116.

Michon, F., Sun, J.J., Kim, C.Y., Ciliberti, D., & Kloosterman, F. (2019). Post-learning hippocampal replay selectively reinforces spatial memory for highly rewarded locations. *Current Biology, 29*, 1436–1444.e5.

Newton, P. M. (2015). The learning styles myth is thriving in higher education. *Frontiers in Psychology, 6*, 1–5. doi: 10.3389/fpsyg.2015.01908

*Painters History of Education (*1886) in Pashler, H., McDaniel, M., Rohrer, D., and Bjork, R. (2009). Learning styles: concepts and evidence. *Psychological Science in the Public Interest, 9*, 105–119. doi: 10.1111/j. 1539–6053.2009.01083.x

Pashler, H., McDaniel, M., Rohrer, D., and Bjork, R. (2009). Learning styles: concepts and evidence. *Psychological Science in the Public Interest, 9*, 105–119. doi: 10.1111/j. 1539–6053.2009.01083.x

Raichle, M. (2015). The brain's default mode network. *Annual Review of Neuroscience, 38, 443–447.*

Reinkensmeyer, D. J., & Patton, J. L. (2009). Can robots help the learning of skilled actions? *Exercise and Sport Sciences Reviews, 37* (1), 43–51. https://DOI: 10.1097/JES.0b013e3181912108

Rink, J.E. (1994). Task presentation in pedagogy. *Quest, 46(3)*, 279–280. https://DOI: 10 .1080/00336297.1994.10484126

Rock, D. & Ringleb, A.H. (2013). *Handbook of Neuroleadership.* CreateSpace Independent Publishing Platform. ISBN-10: 9781483925332

Rogers, F. (2006). *Many ways to say I love you.* Hachette Books. ISBN: 0316492566

Rogowsky, B. A., Calhoun, B. M., & Tallal, P. (2015). Matching learning style to instructional method: effects on comprehension. *Journal of Educational Psychology, 107*, 64–78. doi: 10.1037/a0037478

Rogowsky, B. A., Calhoun, B.M. & Tallal, P. (2020). Providing instruction based on students' learning style preferences does not improve learning. *Frontiers in Psychology, 11*, 1–7. Doi: 10.3389/fpsyg.2020.00164.

Rubin, D.B., Hosman, T., Kelemane, J. N., et al., (2022). Learned motor patterns are replayed in human motor cortex during sleep. *Journal of Neuroscience, 42*(25), 5007–5020.

Ryan, R. M., & Deci, E. L. (2000). Self-determination theory and the facilitation of intrinsic motivation, social development, and well-being. *American Psychologist, 55*(1), 68–78. https://DOI: 10.1037//0003-066x.55.1.68

Schmidt, R.A. & Wulf, G. (1997). Continuous concurrent feedback degrades skill learning: implications for training and simulation. *Human Factors, 39*(4), 509–525. https://doi: 10.1518/001872097778667979.

Schmidt, R.A. & Wrisberg, C.A. (2000). *Motor learning and performance: A problem-based learning approach.* Champaign, IL: Human Kinetics.

van de Ven, G.M., Trouche, S., McNamara, C.G., Allen, K., & Dupret, D. (2016). Hippocampal offline reactivation consolidates recently formed cell assembly patterns during sharp wave-ripples. *Neuron, 92*, 968–974.

von der Kolk, B. (2014). *The body keeps the score. Brain, mind, and body in the healing of trauma.* Viking, ISBN: 0670785938

Willingham, D.T., Hughes, E.M., & Dobolyi, D.G. (2015). The scientific status of learning styles theories. *Teaching of Psychology, 42*, 266–271. Doi:10.1177/0098628315589505

Wong, J. D., Kistemaker, D. A., Chin, A. & Gribble, P. L. (2012). Can proprioceptive training improve motor learning? *Journal Neurophysiology, 108*(12), 3313–3321. https://doi: 10.1152/jn.00122.2012

Wozniak, P. & Gorzelanczyk, E.J. (1994). Optimization of repetition spacing in the practice of learning. *Acta Neurobiologia Experimentalis, 54*(1), 59-62.

Wrzesniewski, A., Schwartz, B., Cong, X., & Kolditz, T. (2014). Multiple types of motives don't multiply the motivation of West Point cadets. *Proceedings of the National Academy of Science, 111*(30) 10990-10995. https://doi.org/10.1073/pnas.1405298111

Wulf, G., & Schmidt, R. A. (1997). Variability of practice and implicit motor learning. *Journal of Experimental Psychology: Learning, Memory, and Cognition, 23*(4), 987–1006.) https://doi.org/10.1037/0278-7393.23.4.987

Yerkes, R.M., & Dodson, J.D. (1908). The relation of strength of stimulus to rapidity of habit-formation. *Journal Comparative Neurology and Psychology, 18*(5), 459–482. https://doi: 10.1002/cne.920180503.

Suggested Readings

Why We Make Mistakes, Joseph T. Hallinan

Between Parent and Child, Dr. Haim G. Ginott

Motor Learning and Performance, Richard A. Schmidt & Craig A. Wrisberg

Tiger's Bond of Power, Chuck Hogan

Coaching With the Brain in Mind, David Rock, & Linda J Page, Ph.D.

The Gamification of Learning and Instruction, Karl M. Kapp

How the Brain Learns, David A. Sousa

The Heart of Coaching, Thomas G. Crane & Leriss N. Patrick

Seeing With the Mind's Eye, Michael Samuels, M.D.

Coaching Questions, Tony Stoltzfu

Free Play, Stephen Nachmamovitch

Rethinking Golf, Chuck Hogan

Shakespeare the Coach, Ric Charlesworth

Thinking and Doing, Moshe Feldenkrais

Subliminal: How Your Unconscious Mind Rules Your Behavior, Leonard Mloinow

The Book, Alan Watts

The Owner's Manual for the Brain, Pierce J. Howard, Ph.D.

Brain Rules, John Medina

Golf From Point A, Susie Meyers & Valerie Lazar

How We Learn to Move, Rob Gray

Thinking Body—Dancing Mind, Chungliang Al Huang, with Jerry Lynch

The Confident Mind, Dr. Nate Zinsser

How Emotions Are Made, Lisa Feldman Barret

The Competitive Buddha, Dr. Jerry Lynch

Small Teaching, James M. Lang

Play Your Best Golf Now, Lynn Marriott & Pia Nilsson

Acknowledgments

Not a day passes without my appreciation for the large number of resources that have directly and indirectly influenced what has been compiled here. I thank everyone who has helped Michael Hebron gain new insights about the nature of learning that I did not have during the first twenty years I was teaching. For the most part, I am sharing what I have learned from others about the brain's connection to learning and memory.

A special thanks to the staff at Harvard University's Connecting the Mind-Brain to Education Institute where I have taken classes and its director Kurt Fischer. I want to thank Eric Jensen, who organized several seminars about the brain's connection to learning that I attended. Dr. Robert Bjork, director of UCLA Learning and Forgetting Lab has mentored and guided me for a number of years, thank you. I see this book as a progress report on how my insights about learning and teaching have changed over time.

The tone of this book was influenced by suggestions and questions from Ryan Hayden, Nicholas Renna, Nannette Poillon-McCoy, Richard Cohen, Professor Stephen Yazulla, Susan Berdoy-Meyers, Fran Pirozzolo Professor of Neuroscience UCLA Medical Center, and Andrea McLoughlin, Ph.D. Associate Professor, Long Island University.

In the mid-1960s Gene Borek, a man I had the highest respect for, suggested that I should write down my impressions of what I was reading, studying, and experiencing. Gene suggested that I write on a scheduled basis—daily, weekly, or monthly; the choice was mine to make as long as I wrote. He felt that if I had a personal record of my thoughts and impressions, I would be able to look back and learn from my own journey of development. Because of Gene Borek, who passed on in 2011, I have been making notes daily for over five decades. Thank you, Gene Borek, you are missed.

Everyone in modern science who has been researching the brain's connection to learning has a big thank you from me. Because of these men and women, we now have a model that can enhance approaches to learning without having to know every aspect of how brains operate.

This model caused me to question the value of some long held cultural assumptions that were not taking into consideration the nature of learning, memory, and emotions that were influencing my approach and methods of instruction for years, which I now realize was A LARGE OVERSIGHT.

__To Be Continued . . . We never know what the future holds, especially when it comes to what may be possible. I know I will keep looking. Best of luck pursuing your goals.__
—Michael Hebron

Michael Hebron School For Learning

Including: Learning Golf Workshops, Seminars on and off Premises,
Private Lessons, and Public Speaking Engagements . . .

CALL: 1-631-979-6534

Online: www.michaelhebron.com
Email: michael@michaelhebron.com

BOOKS BY MICHAEL HEBRON

See and Feel the Inside Move the Outside
By Michael Hebron

The Art and Zen of Learning Golf
By Michael Hebron

Play Golf to Learn Golf
By Michael Hebron

Golf Mind, Golf Body, Golf Swing
By Michael Hebron

Modernizing Approaches to Learning Golf
By Michael Hebron

Learning With the Brain in Mind
By Michael Hebron with Stephen Yazulla Ph.D.

IT DEPENDS!
By Michael Hebron with Stephen Yazulla Ph.D.

SEAN KELLY, Harvard Faculty with focus on various aspects of the philosophical, phenomenological, and cognitive neuroscientific nature of human experience.

"Golf Coaches should look beyond just the golf swing. Mindset and the learning process are important. I agree with you Michael!"

CHRIS COMO, whose career has focused on studying and working with many of golf's preeminent teachers and influencers including Guy Voyer, Stuart McGill, and Dr. Young-Hoo Kwon. Chris has worked with professional Tour players, most notably Tiger Woods and Bryson DeChambeau.

"Michael, I had such a great time chatting with you at the Coaches Camp. Loved going deep with you into the intricacies of instruction and techniques with you but the way you put Mindset at the top of importance is paramount to any long-term success, in my opinion."

GRANT WAITE, Tour victories, Kemper Open, Bell Canadian Open (finishing second to Tiger Woods), New Zealand Open, Trafalgar Capitol Classic, Utah Open, and shot 60 in a final round at the Phoenix Open.

"Michael, I really like how you are seeking out information on how people learn. This is a missing component to golf instruction that is just assumed. Info is what it is, assimilation is the change and is the most important aspect to the student."

Printed in the USA
CPSIA information can be obtained
at www.ICGtesting.com
LVHW081946261023
762119LV00009B/654